The Impact of Hunting on Victoria Crowned Pigeon

(*Goura victoria*: columbidae)

in the Rainforests of Northern Papua, Indonesia

THE IMPACT OF HUNTING
ON VICTORIA CROWNED PIGEON (*Goura victoria*: COLUMBIDAE)
IN THE RAINFORESTS OF NORTHERN PAPUA, INDONESIA

Dissertation
for the award of degree of
"Doctor rerum naturalium" (Dr.rer.nat)
within the doctoral program biology
of the Georg-August University School of Science (GAUSS)

Submitted by
Henderina Josefina Keiluhu
Born in Sumbawa Besar-West Nusa Tenggara, Indonesia

Göttingen, 2013

Bibliografische Information der Deutschen Nationalbibliothek

Die Deutsche Nationalbibliothek verzeichnet diese Publikation in der Deutschen Nationalbibliografie; detaillierte bibliografische Daten sind im Internet über http://dnb.d-nb.de abrufbar.

1. Aufl. - Göttingen : Cuvillier, 2013
   Zugl.: Göttingen, Univ., Diss., 2013

   978-3-95404-567-9

Thesis Committee

Prof. Dr. M. Mühlenberg
Johann Friedrich Blumenbach Institute of Zoology and Anthropology

Prof. Dr. R. Willmann
Johann Friedrich Blumenbach Institute of Zoology and Anthropology

Members of the Examination Board

Reviewer: Prof. Dr. M. Mühlenberg
Johann Friedrich Blumenbach Institute of Zoology and Anthropology

Second Reviewer: Prof. Dr. R. Willmann
Johann Friedrich Blumenbach Institute of Zoology and Anthropology

Further members of the Examination Board

Prof. Dr. C. Leuschner
Albrecht von Haller Institute of Plant Sciences

Prof. Dr. E. Bergmeier
Albrecht von Haller Institute of Plant Sciences

Prof. Dr. H. Behling
Albrecht von Haller Institute of Plant Sciences

PD. Dr. T. Hörnschemeyer
Johann Friedrich Blumenbach Institute of Zoology and Anthropology

Place and date of the oral examination: Computer Room, Department of
Conservation Biology, Center for Nature Conservation, Bürgerstrasse 50, 37073
Goettingen; October 30th, 2013 at 11.15 pm

Acknowledgements

I am very grateful to my supervisor Prof. Dr. M. Mühlenberg, Department of Conservation Biology, Georg-August University of Goettingen for enhancement my concepts about nature conservation. I also thank Prof. Dr. R. Willmann for being my second supervisor, and to Dr. Richard Noske for the valuable tutorial during proposal writing.

The Deutscher Akademischer Austausch Dienst (DAAD) contributed generous financial support for my study. The Head of Education and Teaching Agency of Papua Province on behalf of Governor of Papua Province also provided the financial support for my field research and for finishing my study.

I extend my gratitude and great appreciation to all friends who helped me during this study, thank you very much for all the friendships, supports and togetherness. Particularly, I would like to express my grateful thanks to Dr. M. Pangau-Adam for her advice, assistance and patience.

In this moment, I want to give my gratitude, appreciation and pride to my Mom and Dad (†), my brother (†) and sisters with their families, for always supporting me with love and encouraging me to do the best in my life, these feelings are never enough to say to all of you. Moreover, I extend truthful gratitude for Lewaherila Big Family, for all attention, kindness for being with my children and my husband during this study.

Finally, I express my sincere gratitude to my beloved husband Markus P. Lewaherila, and our children Bryan and Charlie, for their smiles and laughs, their endless love and everlasting encouragement and also for all understanding, sacrifice and praying. I want to say a deep apologize because I already left you all in Papua to pursue a doctorate degree in Germany.

Above all many grateful thanks to my Lord Jesus Christ: Who always gives a way where there seems to be no way.

Henderina J. Keiluhu

Summary

Victoria crowned pigeon (*Goura victoria*) is an endemic bird and has been declared as protected species by Indonesian Government under Law Act No. 301/1991. This species with two other species of *Goura* (crowned pigeon) are endemic to New Guinea islands, and have been state as Restricted Range Species. IUCN Red List also has verified the entitre genus of *Goura* as the largest-body sized of pigeon in the world with status of vulnerable species due to hunting problems, beside listed on Appendix II of CITES as well.

The workshop on Priority-Setting of Biodiversity Conservation in Papua has launched that the major threats on this bird included the large-scale forest conversion for logging, swidden-agriculture, plantation, transmigration, and settlement, also hunting and illegal trading. Local communities in Papua have been practicing hunting on wildlife especially on bird for subsistence, though it is moving towards commercial activities in some regions recently. Since hunting becomes the main threat to *Goura* spp, it is important to conduct a field study on the impact of hunting on Victoria crowned pigeon in tropical rainforests of the northern area in Papua-Indonesia.

The published information on hunting activity and its impact are very limited, as well as the lack data on Victoria crowned pigeon population in its natural habitat in the northern Papua. More over, information on tree communities and vegetation structure in the habitat of Victoria crowned pigeon in this region is still very inadequate. Based on these conditions, it becomes very important to carry out such a research focusing on hunting practice, population of *Goura victoria*, and forest structure in the northern of Papua.

The main aim of this study was to assess the impact of hunting on the population of Victoria crowned pigeon in the rainforests in Papua. The current research is intended to contribute the conservation action of Victoria crowned pigeon in the future. The specific aims of this study includes to investigate forest structure in four different areas inhabited by *G.victoria* and to estimate the population size and density of Victoria crowned pigeon in four different forest areas in northern Papua; The other aims of the study are to compare the population size of *G.victoria* in the given areas; to describe the activity of the bird's hunters and their impact on the population of *G.victoria* in those forest areas and to create and increase awareness of the local people for the conservation of Victoria crowned pigeon.

The study was concentrated in forests of four different regencies in the northern part of the Papua Province, which are forest of Buare (Mamberamo Raya), Supiori (Supiori), Unurumguay (Jayapura), and Bonggo forest (Sarmi). The detailed observations on population density on *Goura victoria*, composition and forest structure, also on hunting activity by local people, and its impacts on the population of Victoria crowned pigeon were conducted in those forest areas. Buare and Supiori forests are the parts of nature reserve become forest area with lower interference of local people activity compared with higher interference and pressures in Unurumguay and Bonggo forests.

In each study site, as many as 25 randomly long lines transects for vegetation analysis of 20 x 100 meters with 20 m x 20 m plots were established for vegetation analysis. Measurement and identification within each plot were taken on each tree with a diameter at breast high more than 10 cm and more than a meter height. Furthermore, floristic structure was assessed quantitatively by calculating the Important Value Index (IVI) for each species in each study site. The IVI represents the sums of the value of Relative Density (RD), Relative frequency (RF) and Relative Dominance (RDo). *Goura* surveys were carried out at four sites using line transect method and 45 transects were set aside in all study sites. The researcher walked along the transect line and recorded the perpendicular distance between detection points and transect line. Surveys were done four days per week, between 06.00 in the morning to 16.00 in the afternoon each day by the field team. The semi structural interviews with questionnaires were used and the interviews were conducted on 151 respondents who live in 13 villages of four districts in four regencies. Important Value Index (IVI) and Shannon-Wiener Diversity Index (H') were used to calculate the floristic composition and forest structure in each study site. Distance 5.0 release 2.0 program was used to estimate the population density and population of Victoria crowned pigeon. The Mann Whitney U test, Kruskall-Wallis test and Multiple Linier Regression Analysis using SPSS version 19.00 were used to illustrate the hunting activity by local people and predict the impact of hunting by local people to *Goura victoria* population. Then, the estimation of maximum sustainable annual harvest was compared to the value of maximum current annual harvest. Hunting practice on *Goura victoria* is unsustainable if the value of maximum current annual harvest exceeds the value of maximum sustainable annual harvest. All data analysis was processed using Excel program.

Floristc composition in each study sites showed that the 58 species in 38 families in Buare and 57 species in 38 families were found in Supiori. These were quite different with 39 species in 25 families found in Unurumguay and 34 species in 22 families found in Bonggo, The tree diversity in each study site also varied, showed as H'= 3.55 in Buare forest, 3.45 in Supiori, 3.09 in Unurumguay and 3,00 in Bonggo. Although the diversity in Buare seems more diverse than that in other sites, it is statistically not significant, because the values of H' of all study sites are in the range between 1– 4.5. The seven most dominant tree species based on the Important Value Index were varied between study sites. These species belong to different families, with Euphorbiaceae family as the most common family encountered in all study sites. The results showed that *Pimeliodendron amboinicum* Hassk become dominant tree species in forest area of Buare, Supiori and Unurumguay, while *Pometia* spp. (*Pometia pinnata* and *Pometia* sp.) dominated forest area in Buare, Unurumguay and Bonggo. Likewise, the measurements of diameter at breast height and tree height class distribution were used to describe structural composition of forest area in each study site. This result shows that about 80% of vegetation in all study sites was represented by trees with diameter at breast height less than 30 cm and Bonggo area has trees with small diameter and already loss the large trees. Trees from all diameter class in other three locations had descending trend quantity from small to big diameter, while all study sites showed similar forest structure in distribution of trees height.

Population size of *Goura victoria* was varied, which depends on the size of hunting area with higher value of estimation on population density but has the least value population size of *Goura victoria*.

The interviews with Papuan hunters about hunting practices showed that distance of hunting area, hunting using air gun, using dogs and using foot snares on hunting Victoria crowned pigeon were varied among each study sites. Hunters in Buare area prefer to hunt *Goura victoria* within the distance of less than 2 - 5 km, while mostly hunters in three other sites prefer to hunt *Goura victoria* within the distance of 3 km to more than 5 km. The used of air gun in Buare area, was not recorded, while in the three other study sites it was more common though in low level, not more than 22% of all hunting practices. Using dogs in hunting *Goura victoria* also occurred less frequent in all study sites, only about 12%. Hunters in all study sites tended to use foot snares in catching *G.victoria*. However, the estimation value of current annual harvest within the hunting area size for each study sites exceed from the allowable values on estimation of maximum sustainable annual harvest per each hunting area size.

Goura's hunting is already prohibited not only in Indonesia, but also in Papua New Guinea. *Goura victoria* as a high-valued bird, is mainly sold alive and usually being hunted for fresh money to fulfill daily needs of hunter's family.

Hunting activities in all study sites were relatively high compared to other areas of Papua, an example from hunting of *G.victoria* in Waropen showed the high frequency of hunting activity. Hunting activity on Victoria crowned pigeon's was unsustainable and this practice by local people has negative effects of *G.victoria* population, although most of hunters using foot snares.

The result from ths study showed that protection of Victoria crowned pigeon needs deep concern from the Governments. The related stakeholders should enhance and determine conservation areas with the factual boundaries, including protected forest, animal sanctuaries and nature reserves. It also necessary to establish and manage more buffer zones around protected area immediately, to reduce interference from local people. Papuan people need more socialization of the laws and regulations concerning wildlife protection. The law enforcements should be implemented together with strict sanctions. Further research on *Goura victoria* should be carried out on other part of northern Papua, including short and long terms in all ecological aspects of *Goura victoria*.

TABLE OF CONTENTS

List of Figures

List of Appendices

CHAPTER 1: INTRODUCTION

1.1. Background

Indonesia is known as one of the richest biodiversity countries in the world, with the most complex ecosystems (Petocz 1987). The country covers only 1.3% area of the globe, but it possesses about 10% species of flowering plants, 12% of mammals, 16% of reptiles and amphibians, 17% of birds and 35% of fishes from whole species in the world (BAPPENAS 2003).

New Guinea Island is the second largest island in the world and known as the largest between all tropical islands. This island is accounted amongst the richest biodiversity areas and has the most diverse assemblage of ecosystems on the earth. New Guinea consists of one big island with some small satellite islands, and administratively belongs to Papua-Indonesia and Papua New Guinea (mentioned further as PNG). Papua is biogeographically a part of Melanesia region and has already been stated as priority tropical forest area. Conservation International (1997) declared New Guinea as "Major Tropical Wilderness Area" (Supriatna 1997). Concerning bird diversity, New Guinea has approximately 831 species or represents around 8.6% of the birds in the world, while Papua, Indonesia has approximately 657 bird species or 6.8% of the total birds in the world (Mack and Dumbacher 2007), and 25% of the total bird species in Indonesia (Petocz 1987). In addition, Bird Life International has been identified about 140 Endemic Bird Area's (EBA) worldwide, and eight of EBA sites were located in Papua (Sudjatnika *et al* 1995). Furthermore, Mack and Dumbacher (2007) stated that family Columbidae in Papua has the richest species rate, 42 species out of 309 species worldwide.

The percentage of endemic birds in Papua is higher than in other areas in Indonesia, and these birds are mainly dispersed into five nature conservation areas (Petocz, 1987). These conservation areas include Arfak Mountains with a total of 278 bird species, Tamrau Mountains 146 species, Lorentz Mountains 130 species, Mamberamo region 191 species, and Wasur areas 74 species.

Papua (formerly Irian Jaya) is the largest, but least-developed province of Indonesia. Although there is a great lack of biological data for Papua compared with the other half-eastern part of New Guinea (PNG), it has been estimated that the province has around 50% of Indonesia's biodiversity (Supriatna 1997, 2008).

Until 1984 Papua-Indonesia had escaped the devastating extent of deforestation that strikes other part of Indonesia and South-East Asia (Anggraeni 2007). However, the forest area in Papua decreased in 1993-1997 from 90% to 80% of the total area of the Island (Supriatna 1997). Moreover, the rate of forest conversion increases constantly since Indonesia's recent economic crisis (Richards and Suryadi 2002). Nowadays, loss of habitats and forest fragmentations due to logging, plantation, transmigration, cultivation, mining, oil and gas extractions, and rapid development of settlements and roads result in threats on Papua's unique biological heritage (Richards and Suryadi 2002). Other development projects such as the expansion of oil palm plantations become a seriously menace to the existence of tropical lowland forest in Papua (Smolker *et al* 2008, Samuelson 2008).

Papua is also a home to more than 250 different ethnic groups, each with their own rich culture, tradition, language and sets of interrelationships with their environment (Petocz 1987, Supriatna 2008). Indigenous people have already a long history of subsistence in hunting, fishing and cultivation systems. Shifting cultivation system has occurred for about 5,000 years in Papua and hunting activity has known since 3,500 years ago (Hope 2007). People living in the lowlands and swamp areas traditionally rely heavily on sago, while highland people practice rotational cultivation system of bulb and root crops mainly on taro and sweet potatoes. Pigs are raised as source of protein, but additionally the people preferred hunting wild pigs, along with other wild animals from the forest (Petocz 1987, Boissière *et al* 2007).

Hunting is a major activity of indigenous people in Papua, yet there is no quantified studies undertaken on the impact of hunting on wild animal. Usually hunting is practiced in subsistence manner, and commercial hunting occurs only at a small scale or in the heavily capitalized region (Bennett and Robinson 2000[a]). Hunting might also be carried out only for cultural occasion or recreational reasons. The meat from hunted animals could be distributed within the community or might be sold in local marketplace (Pangau-Adam and Noske 2010, Pangau-

Adam *et al* 2012). In many hunting studies, however, the distinctions between hunting for subsistence and for commercial purpose are rarely clear (Dwyer and Minegal 1991, Pangau-Adam and Noske 2010, Pangau-Adam *et al* 2012, Aiyadurai *et al* 2010, Aiyadurai 2011, Bennett and Robinson 2000[a], Lee 2000, O'Brien and Kinnaird 2000). Bennett and Robinson (2000[a]) and Mansoben (2005) stated that it is essential to understand the cultural and socioeconomic context, and to collect accurate information on hunting and its effects, in order to determine the sustainability of this practice.

Ground-dwelling crowned pigeon (*Goura victoria*) is an endemic bird species which is declared as protected species by Indonesian Government (Law Act No. 301/1991). All three species of *Goura* (crowned pigeon) are on the status of Restricted Range Species and endemic to New Guinea and its satellite islands (Rand 1938, Beehler *et al* 1986, Andrew 1992). Furthermore, the entire genus *Goura* as the largest pigeon in the worlds has been verified by IUCN Red List as vulnerable species due to hunting problems (Collar *et al*, 1994; IUCN 2011), and also listed on Appendix II of CITES (Statterfields *et al*, 1998).

The workshop on Priority-Setting of Biodiversity Conservation in Papua held in 1997 has founded that the major threats on this bird were the large-scale forest conversion for logging, swidden agriculture and plantation, transmigration, settlement, hunting and illegal trading (Supriatna 1997). Hunting on wildlife especially bird species in Papua has been practiced by the local communities for subsistence, but in some regions it is recently moving towards the commercial activities (Pangau-Adam, 2010, Suryadi *et al* 2007, Mahuse 2006, Sada 2005). Because hunting is amongst the main threats to *Goura* spp, it is due importance to conduct the field study on the impact of hunting on *Goura victoria* in the tropical lowland rainforests of the northern Papua-Indonesia.

1.2. Objectives of the study

There are only three species in the genus *Goura* (crowned pigeons) and all are endemic to New Guinea. *Goura cristata* inhabits lowland area of the Bird's head and Bird's neck, *Goura scheepmakeri* inhabits southern lowlands, and *Goura victoria* inhabits the lowland areas of the northern New Guinea (see the distribution map in figure 1.1) below.

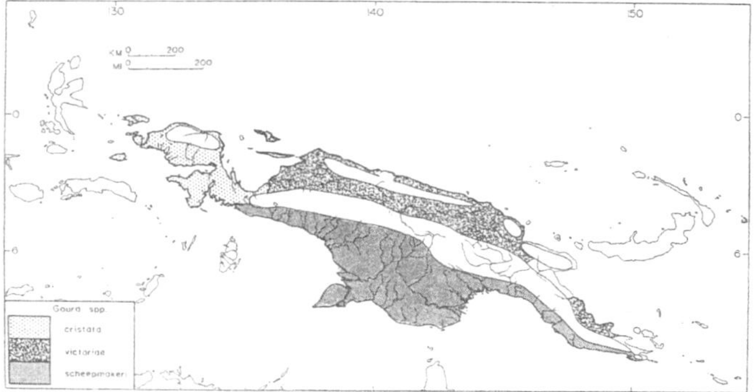

Figure 1.1 Distribution map of three species of Genus *Goura* in
New Guinea (Source: Pratt 1982)

All *Goura* species play an important role in the traditions and daily life of the indigenous people in Papua. These birds are among the target animals of wildlife hunting. There is still very little published data on the hunting activity and its impact, as well as lack of data on *Goura* population in its natural habitat in the northern Papua. The only reports on *Goura* population are those from King and Nijboer (1994) and Bird Life International (2012). On the other hand, information on tree communities and vegetation structure of *Goura* habitat in this region is still very limited, although a rapid assessment had been conducted in Mamberamo area (Richards and Suryadi 2002). Based on these considerations, it is important to conduct such a research with focus on hunting practice, population of *G.victoria*, and forest structure in the northern Papua.

The main aim of the study is to assess the impact of hunting on the population of *G.victoria* in the lowland rainforests in Papua. The current research is intended to contribute into conservation action of *G.victoria* in the future.

The specific aims of this study include (1) To investigate the forest structure in four different areas inhabited by *G.victoria*; (2) To estimate the population size and density of *G.victoria* in four different forest areas in northern Papua; (3) To compare the population size of *G.victoria* in four different forest areas; (4) To describe the activity of the bird's hunters and their impact on the population of

G.victoria in four different areas and (5) To create and increase awareness of the Papuan people to conserve *G.victoria.*

In the last two decades, the increase of forest degradation in the northern Papua occurred due to the people activities such as illegal logging, illegal wildlife hunting, collecting and trading forest products (Jepson *et al* 2011, Suryadi *et al* 2007). The Special Autonomy Laws ratified in 2001 has lead the natural utilization to increased forest conversion into various development purposes, such as new districts, regencies, roads, resettlements, and the establishment of vast area for oil-palm plantations (Smolker *et al*, 2008; Samuelson, 2008). Furthermore, illegal logging and other forest conversion may facilitate hunters to reach the remote forest areas (Wilkie 1989, Kinnaird *et al* 2003, Miranda *et al* 2003, Suryadi *et al* 2007, Frazier 2007)

Studies about wildlife hunting have been carried out in some tropical regions in the world. In African forest for instance, individual hunter primarily hunt the wild animals to eat and sell the captured animals (van Vlieth and Nasi 2008). Mammals have become the main source of bushmeat protein throughout Africa (Fa and Brown 2009), for example hunting on duikers in Guinea, West-Central Africa (Pailler *et al* 2009). Moreover birds are also hunted in other parts of Africa for sport, cash or subsistence (Waltert *et al* 2010, Thiollay 2005, Hart and Upoki 1997).   In the Neotropic regions, hunters commonly harvest many species of wildlife animal like tapir, brocket deer, armadillos, agoutis and several species of birds (Bodmer *et al* 1995 and 1997, Peres 2000, Mena *et al* 2000). The Amazonian hunters also hunted birds, especially the big-size birds such as Great Tinamous, Great Curassow and Crested Guan (Smith 2005, Peres 2000, Mena *et al* 2000, Begazo and Bodmer 1998). In Indonesia, hunting of wildlife animals has been widely noted, for instance hunting on Bornean peacock-pheasant in Borneo (O'Brien *et al* 1998) and large birds and mammals in the North Sulawesi (O'Brien and Kinnaird 1996 and 2000, Lee 2000, Alvard 2000). Related to sustainable hunting, current subsistence hunting and commercial hunting in different tropical regions are tend to be unsustainable (Noss 2000, Begazo and Bodmer 1998, O'Brien and Kinnaird 2000, Lee 2000, Alvard 2000, Robinson and Redford 1994).

Deforestation along with hunting practice raised some critical questions related to the sustainability: how is the current population of *G.victoria* in lowland forest of the northern Papua? Can *G.victoria* survive in different habitats? Does the traditional hunting for wild meat consumption lead to reduce *Goura* population? In order to answer these questions, it is very important to study the population density, forest habitat condition and hunting on the *G.victoria*.

The first question about the population *of G.victoria* emphasizes that we have insufficient data and information on the current population size of this species. The data on population status of *Goura* is very limited, and the only record stated that there are about 2,500 to 9,999 individuals inhabit the lowland forest in the northern Papua, with decreasing population trend (Bird Life International 2012). Headed for answering the question and obtaining preliminary data about population density, the research on measuring *Goura* population become the right choice.

Second question is with regards to the habitat condition of *G.victoria*. This bird species needs undisturbed habitat for nesting, foraging and breeding, but currently habitat disturbance and forest clearing for human needs are threatening the persistence of *G.victoria*. Reduction in forest area is leading to the decline and degradation of habitat area, and undoubtedly affects the population of the bird. These activities should be concerned whether it might have a negative impact on *Goura's* population size or not.

The third question is food gathering and hunting activities in the lowland forest. These are two activities that have been common for the traditional forest-dwellers in Papua. *Goura victoria* has become one favorable source of wild meat for the nutrition of hunter family (King and Nijboer 1994). Additionaly local people also use the feathers of *G.victoria* as a head decoration for traditional Papuan dancer (King and Nijboer 1994, Pattiselano and Mentansang 2010).

The entire questions and facts mentioned above are leading to the important issue concerning urgent conservation efforts for the endemic bird species, *G.victoria,* in lowland forest of the northern Papua. Forest degradation might frequently occur in the lowland forest, and also towards the forest reserves and wildlife sanctuaries containing lowland forest. Habitat degradation and unsustainable bird hunting may threaten the existence and persistence of population of *G. victoria*. It is expected that the information on population density, combined with an analysis on habitat that utilized by this species, and the information on traditional hunting activities can lead to the comprehensive output about conservation status of *G. victoria* in Papua.

CHAPTER 2: GENERAL REVIEW ON *Goura victoria*

2.1. Biology of Genus *Goura*

2.1.1. Systematics of the Genus *Goura*

Order of Columbiformes is composed by three families: the sandgrouse (Pteroclidadae), the dodos (Rhapidae), and the pigeons (Columbidae), but Family Rhapinae was already extinct during the 17-18th centuries (Harrison *in* Nijboer & Damen 2000). Family Columbidae consists of 5 subfamilies with 42 genera, 749 taxa and about 309 species totally (Baptista *et al* 1997). Gibbs *et al* (2001) discovered that Columbidae consists of 5 families with 42 genera and around 316 species, slightly different from Beehler *et al* (1986) which stated that Columbidae consists of around 299 species.

The species of Columbidae are distributed widely and can be found all over the world except in polar and sub-polar regions, in extremely hot and cold regions, and in dome oceanic islands. The term of Columbidae sometimes is used to characterize the birds based on the similarity of their size, typology and ecology, but it is inconsistently used and not based on any real biological dissimilarity (Goodwin 1983 and Beehler *et al* 1986). Gibbs *et al* (2001) classified pigeon and dove based on the size. Pigeon generally refers to the larger species while dove to the smaller and more elegant species. Additionally, the term pigeon and dove are somewhat interchangeable. Both groups are unique among other birds in Columbidae due to their production of "crop milk" that is secreted by sloughing of fluid-filled cells from their crop layer (Perrins 2009, Baptista *et al* 1997). In these groups, both male and female can produce this highly nutritious substance to feed their juveniles (Beehler *et al* 1986: Baptista *et al* 1997).

The Gourinae is one of the subfamilies in Columbidae that contains only the three species of Crowned Pigeons. The other subfamilies are Columbinae (the typical seed-eating pigeons), Treroninae (the fruit-eating pigeons and fruit-eating doves), Otidiphabinae (the pheasant pigeon), and Didunculinae (the tooth-billed pigeon) as the largest subfamily among the order of Columbiformes (Goodwin, 1983, Baptista *et al* 1997 and Gibbs *et al* 2001). Sub-family Gourinae consist of one genera, and genus *Goura* comprises three species, *Goura cristata, Goura scheepmakeri* and *Goura victoria* (Beehler *et al* 1986, Baptitsta *et al* 1997, Nijboer and Damen 2000, Gibbs *et al* 2001). Every species consists of two sub-species, *Goura cristata cristata*, Pallas 1764; *Goura cristata minor*, Schlegel 1864; *Goura scheepmakeri scheepmakeri*, Finch 1876; *Goura scheepmakeri sclaterii*, Salvadori 1876; *Goura victoria victoria*, Fraser 1876 and *Goura victoria beccarii*, Salvadori 1876.

2.1.2. The Distribution of Genus *Goura*

All species of Crowned Pigeons are similar and geographically interchangeable each other (Figure 2.1). The three species are also very closely related, and inhabit only in New Guinea and its satellite islands (Peckover and Filewood 1976; Beehler *et al* 1986). Their distribution is mainly allopatric, but two of this species (*G.cristata* and *G.victoria*) usually meet and hybridize naturally in the Siriwo River at the tip of Cenderawasih Bay on the north-west of New Guinea (figure 2.1 and figure 2.2, Beehler *et al* 1986; Goodwin, 1977; Baptista *et al* 1997). In regard to the distribution and evolution, there is a theory stated that many rainforest birds was the product from isolation of forest refugees during the Pleistocene and post-Pleistocene era (Haffer 1969 and 1974 *in* Pratt 1982). This theory also emphasized that the isolation in remnant forest tracts had divided the widespread population of forest birds into discrete fragmentary populations (Haffer 1969 and 1974 *in* Pratt 1982). Some of these populations were then be able to differentiate as new subspecies or species (Mayr 1963 *in* Pratt 1982).

Figure 2.1 *Goura cristata*, *Goura victoria* and *Goura scheepmakeri* (Note: from left to the right, source: Coates and Peckover, 2001)

Additionally, in the case of Crowned pigeon, the distribution might become evidence of the distribution of allopatric and parapatric species, and also as the result of ecological compatibility and geographic isolation, though their range might be similar to the case of parapatric species (Haffer 1969 and 1974 *in* Pratt 1982).

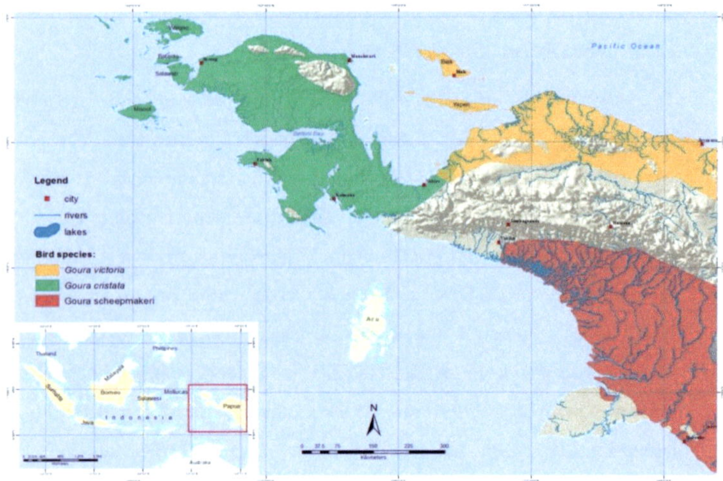

Figure 2.2 Distribution map of three species of genus *Goura* in Papua-Indonesia
(Created by H. Suhendy *base on* Birdlife 2001)

In particular, *G.cristata* inhabits flat lowland forest, usually in undisturbed alluvial forests (Beehler *et al* 1986). These area including the marshes and seasonal flooded area from the north western of New Guinea, until the Etna Bay (on the west of New Guinea' south coast) to the point where the Siriwo Rivers flow into the Geelvink Bay (at the coast in the north of New Guinea) (Rand and Giliard 1967). This part is called the Vogelkop or formerly called Arfak or Berau Peninsula (Gyldenstolpe 1956 *in* Nijboer and Damen 2000), and also Onin Peninsula. This was the area where *G.cristata* was detected hybridized with the *G.victoria* (Beehler *et al* 1986). Furthermore, *G.cristata* also is recorded at some islands close to the coast, like Misool, Salawati, Batanta and Waigeo Island (Rand and Gilliard, 1967; Beehler *et al* 1986; King and Nijboer 1994) and Seram Islands, Moluccas where it could probably be imported (Kitchener *et al* 1993).

Goura cristata can be found from sea level to around 110-150 above sea level (Beehler *et al* 1986, Baptista *et al* 1997). This species was also reported from the Moluccas Islands even it was considered as exotic to the islands (Iridale 1956). In spite of its exotic condition, the first record on this species by Pallas (1844) had specifically showed that the species came from Banda Island, Moluccas (Rothschild 1931 *in* Nijboer and Damen 2000). The size of this species in Moluccas Islands is generally smaller than its conspecifics from the mainland of Papua. Due to this size difference, some scientists distinguished the *G.cristata cristata* on the mainland with the *G.cristata minor* from some islands (described by Schlegel *in* Rand and Gilliard, 1967). On the Misool Island, northwest Papua, an even smaller sub-species named *G.cristata pygmaea* was recorded as well (Mees 1965 *in* Nijboer and Damen 2000).

The second species in genus *Goura* is *Goura victoria*. This species has two subspecies; *G.victoria victoria* which its nominated form lives on the Biak Islands and Yapen Island (formerly called Jobi Island) although it might be introduced to the later (Rand and Gilliard 1967), and *Goura victoria beccarii,* which was named after an Italian explorer Beccarii, similar with the name of a hybrid from the Victoria Crowned Pigeon and Common Crowned Pigeon (Iridale, 1956). This subspecies can be found in forests on the mainland of the northern New Guinea, from Siriwo River (Geelvink Bay) in the west to the Astrolabe Bay or Milne Bay in the east (Beehler *et al* 1986), and at the western end of its range. It overlaps with the distribution of smaller Common Crowned Pigeon, *Goura cristata* (Peckover and Filewood 1976). *G.victoria beccarii* occupies swamps of sago forest and drier forests, found particularly in lowlands, but sometimes it might occurred up to 400-600 m above sea level like at Jimmy Valley (Baptista *et al* 1997). Another theory showed that this bird can be found at the nearby sea level only (Beehler *et al* 1986).

The third species of genus *Goura* is *Goura scheepmakeri* or Scheepmaker's Crowned Pigeon. In body size, this is the largest species of crowned pigeons, with the height around 71-79 cm and weight about 2000-2235 g (Baptista *et al* 1997). The two subspecies of this species are the Hall Sound (*Goura scheepmakeri scheepmakeri*) and the sub-ordinate *Goura scheepmakeri sclaterii.*

The Hall Sound is distributed on Orengerie Bay (south-eastern part of New Guinea) and the sub-ordinate can be found between Mimika River and Fly River in the South of New Guinea (King and Nijboer 1994).   The range of *G. scheepmakeri* might be extended until Etna Bay (Beehler *et al* 1986), however it has not been recorded whether and where the two sub-species meet (Peckover and Filewood 1976).

2.2. General Morphology of Genus *Goura*

2.2.1. The Common Crowned Pigeon (*Goura cristata* Pallas 1764)

The Common Crowned Pigeon (*Goura cristata*) has many names, such as Western Crowned-Pigeon, Masked Goura, Masked Crowned-Pigeon, Grey Crowned-Pigeon, Blue Goura, Grey Goura (English), Goura couronne (French), Gura occidental (Spain) and in Germany as Krontaube (Beehler *et al* 1986; Goodwin, 1983; and Baptista *et al* 1997).

In Papua Indonesia, this bird is recognized locally as 'mambruk polos' or 'mambruk kelabu', and also as 'mambruk Ubiaat' (Beehler *et al* 1986). This bird was discovered by Dampier in 1700 (Iridale 1956) and was firstly described scientifically by Pallas 1764 as *Columba cristata* (*see* Mc Moris 1976 *in* Nijboer and Damen 2000). Other names of this bird were *G.coronata* (Linnaeus *in* Nijboer & Damen 2000) and *Goura cinerea* (Hartert 1895 *in* Mayr, 1941), but "*G.cristata*" is generally accepted recently.

Besides body features that are mentioned earlier, Common Crowned Pigeon also has a large blue "crown" on the head. Each of its wings has a small white spotted mark. This species is blue-greyish, with some paler grey part or creamy tinge on its breast. As for all *Goura*, this species has a well-built body with rather long and stout legs, equipped with a larger laterally-compressed crest of lacy feathers (Nijboer and Damen 2000). The upper part of the mantle and most wings are dark purplish red or dark wine red. This species is infinitely a subject to partial melanism and the individuals with varying or often extensive black where patch on around the head, back to upper tail-converts and belly to under tail-converts (Gibbs *et al* 2001). Melanism is the condition of increase on black or nearly black pigmentation of the feathers that seems to occur more frequently in

G.cristata minor than *G. cristata cristata* (Goodwin, 1977). The Common Crowned Pigeon usually have clown-like patterns in their body, but some others are all greyish. This bird has around 66 cm height (Beehler *et al* 1986) and 1800-2400 g weight (Baptista *et al* 1997).

2.2.2. The Victoria's crowned pigeon (*Goura victoria* Fraser 1844)

Victoria's Crowned Pigeon was described firstly as *Lophyrus victoria* (Fraser 1844 *in* Proc. Zool. Soc. London, page 136 *in* Mayr 1941 and Nijboer & Damen 2000). This bird has several different names such as White-tipped Goura, White-tipped Crowned Pigeon (English), Goura de Victoria (French), Fächertaube (Germany) and Gura Victoria (Spain) (Goodwin 1977; Baptista *et al*, 1997). In Papua Indonesia, this bird is called locally as 'mambruk raja' or 'mambruk kembang". The bird has two subspecies, *G.victoria victoria* (described by Fraser 1984) and *G.victoria beccarii* (described by Salvadori 1876). The name of "Victoria" has been given to the bird as an honour to Queen Victoria, England's queen at that time (Fleay 1961 *in* Rand and Gilliard 1967).

This species can be distinguished simply from the other two *Goura* species due to its white tips on the crest. The crest of Victoria's Crowned Pigeon is blue with combination between blue and white tips and the barbs at their ends are only slightly separated (Baptista *et al* 1997). This bird is darker than the Common Crowned Pigeon. Furthermore, Victoria's Crowned Pigeon has a pale blue spot on each wing that are very-well visible if the bird is not spreading its wings. Actually, the bird's general colour is dark-greyish blue, with dark-purplish red breast, the wings are patched pale-greyish blue, with dark purple edges, the irises are red or purplish red, the beak is dark grey, and the bird has purplish red legs and feet (Baptista *et al* 1997). The bird's nominated form is slightly smaller and rather darker in colour. Similar with Common Crowned Pigeon, Victoria Crowned Pigeon is about 66 cm in size (Beehler *et al* 1986), with about 2000 g in weight (Baptista *et al* 1997). The biggest individual recorded was 74 cm height and weight of 2384 g (Baptista *et al* 1997).

The main diets of these species usually consist of fallen fruits from forest trees, including berries and hard-coated seeds (Peckover and Filewood 1976,

Coates 1985). In captivity the birds can be adapted to sliced fruit, grapes, lettuce, maize, carrots, peanuts and especially fond of the wild fig fruit *Ficus macrophylla* (Fleay 1961 *in* Rand and Gilliard 1967)

2.2.3. The Scheepmakeri's crowned pigeon (*Goura scheepmakeri* Finch 1876)

This species was discovered by the Italian explorer D'Albertis and firstly described by Finch (1st of April 1876 *in* Proc. Zool. Soc. London 1875 page 631 plate 68 *in* Mayr 1941 and Nijboer & Damen 2000). *Goura scheepmakeri* is also named as "Maroon-breasted Crowned Pigeon, The Southern-Crowned Pigeon, Scalter's Crowned Pigeon, Scheepmaker's Crowned Pigeon and Great Goura (English), Goura de scheepmaker (French), Gura Surena (Spain)" and in Germany as "Maronenbrust-Krontaube" (Goodwin 1977, Beehler *et al* 1986 and Baptista *et al* 1997). In Indonesia, this species is named locally as "mambruk besar" or "mambruk ungu".

The Scheepmakeri Crowned Pigeon is just as blue as the Common Crowned Pigeon, but its blue colour is more intensive than that of the Victoria Crowned Pigeon. The Scheepmakeri has a deeply red breast, its crest only has blue colour, and the wings have a brightly white spot, which is larger if compared with the spot on Common Crowned Pigeon wings. The Scheepmakeri Crowned Pigeon is differed from Common Crown Pigeon due to its colour on certain body parts including the dark purplish red belly and breast specifically below the neck (Gibbs *et al* 2001, Baptista *et al* 1997). Its mantle and smaller wings are covered with dark-greyish blue feathers like upper part of the breast and the wings are patched with very pale whitish grey.

The subspecies *Goura scheepmakeri scheepmakeri* is slightly different from the other sub species *G.scheepmakeri scalaterii*. The lower breast and belly of *G.scheepmakeri sclaterii* are greyish blue but in *G.scheepmakeri scheepmakeri* dark purplish red (Baptista *et al* 1997). The breast and the belly of both sub species can be in maroon colour like that of Victoria Crowned Pigeon (Beehler *et al* 1986). Scheepmakeri's irises are deep red and the bill is dark bluish grey and it has purplish red legs and feet (Baptista *et al* 1997). Partial melanism also occurs in Scheepmakeri Crowned Pigeon, but not as frequent as in the Common Crowned Pigeons.

Originally, Scheepmakeri Crowned Pigeons live along the whole south coast of New Guinea, but according to Beehler *et al* (1986) the range of *G. scheepmakeri* probably extends until Etna Bay. *Goura scheepmakeri scheepmakeri* inhabits the dry and flooded lowland forest from Hall Sound to Orangerie Bay at the south eastern part of New Guinea (Rand and Gilliard 1967). The other subspecies *G.scheepmakeri sclaterii* can be found between Mimika River and Fly River in the south of New Guinea (King and Nijboer, 1994). These two sub-species usually live separately in long distance and this might be the reason why there are many variations between them. It seems that *G. scheepmakeri* is fully disappeared from south-eastern part, as the result of the increase of human population in the area, but this species can be reasonably safe in the west of New Guinea area that has less human population (Beehler *et al* 1986).

2.3. Literature review on *Goura* ecology

The information on biological aspects of the three *Goura* species is insufficiently available. It has been stated that they can move in small groups of 2-10 birds, although flocks of up to 30 have been reported (Coates 1985). These birds spend much time foraging on the ground, but the resting and roosting usually on trees (Rand and Gilliard 1967, Coates 1985). They might be wide-ranging and erratic in their movements (Beehler 1982). These birds' common diet consists of fallen fruits, seeds and berries (Coates 1985), and they are also referred as seed predators (Beehler 1982). Crowned pigeons are reportedly attracted to refuse at sago palm preparation site (Beehler *et al* 1986), and *G.scheepmakeri* has also been observed feeding on small crabs (Baptista *et al* 1997, Gibbs *et al* 2001).

The observation in captivity showed that *Goura* lays one egg, which is incubated for 28-30 days (Nijboer and Damen 2000, Beltermann and Poots 2008). The juveniles usually leave from the nest at 28-36 days of age, at which time they are roughly one-third to one-half of mature size.

Full-mature size is reached only after three months, but the juveniles still depend on their parent for several months (King and Nijboer 1994). Another research showed that captive-reared *Goura* of both sexes can be reproduced successfully when it reached the age of 15-17 months old (King and Nijboer 1994). This result could be different compared with the wild population, which age of the first reproduction varies. The evidence from captivity also suggested that *Goura* are relatively slow in their reproduction and development, where data from wild populations are still insufficiently available (King and Nijboer 1994).

Originally, all crowned pigeons are considered common over their range, but recently they are absent from large areas due to hunting pressure, and hence can be found numerous in remote areas only (Rand and Gilliard 1967), especially *Goura victoria*. *G.victoria* is primarily killed for meat, although its feathers are sometimes used for head-dresses as well (Coates 1985). Its eggs and hatchlings are also taken to be reared for food (Birdlife International 2012), and this bird is quickly hunted by hunters in any forests within a day's walk from the village (Beehler1991 *in* King and Nijboer 1994). This species is usually extirpated from the forests around transmigration settlements because the birds are intensively exploited by transmigration settlers, but it might survive from hunting by native people (King and Nijboer 1994).

Basically, the nature and behaviour of crowned pigeons make them particularly susceptible to hunting pressure (King and Nijboer 1994). If being disturbed, they prefer to walk or run away to the remote area, but in demanding situation, this bird will fly noisy to the high branches where they may balance themselves clumsily and gawk at the intruder, and making easy targets of themselves (Rand and Gilliard 1967). Additionally the *G.scheepmakeri* was remarked as "stupidly tame bird" (Bell 1977 *in* King and Nijboer 1994), due to its behaviour. Large concentrations of *G. cristata* have been observed at waterholes in West Papua where they could be easily netted. It remains to be determined whether *G.victoria* exhibits similar behaviour. Based on recent information, these birds are under threats due to wildlife trade and logging in their lowland habitat, although the hunting on *G.victoria* is well justified (Bird Life International 2001, Suryadi *et al* 2007). Genus *Goura* is highly prized as an aviary bird by bird parks, zoos and private aviculturists throughout the world. Despite national protection legislation and listing on CITES Appendix II, Indonesian CITES authorities recorded an export of 200 specimens of *G.scheepmakeri* from Merauke in 1992 (King and Nijboer 1994). It is stated that comprehensive evidence on the number of exported pigeons from Papua, which was recorded to CITES by animal handlers in Singapore, was under the true value (King and Nijboer 1994). A large number of Crowned Pigeons was rumoured being sent illegally. For instance, about 560 Crowned Pigeons were taken away from a feeder-handler in Amsterdam on 1991 (King and Nijboer 1994). Accordingly, there is an urgent need to determine the trade level of all *Goura* species.

It has been argued that captive breeding cannot supply the demand for *Goura* as aviary birds due to its low reproduction success (King and Nijboer 1994). During 1988-1990, populations of *Goura* spp in the North American and European studbook suffered from a negative population growth collectively (Nijboer and Damen 2000). This situation was similar in Southeast Asia as well. It was also reported that much effort was being invested into developing techniques to improve breeding success, so that the situation might be different recently (King and Nijboer 1994).

In Papua, the research on this genus is obviously under developed in their natural habitats, even though several researches have been conducted (Supriatna 1997). More research about *Goura* species were already carried out in zoos including in the bird parks and safari parks (Handini *et al* 1992, Setio *et al* 1996, Roembino 1997). As the results, the most data and information about *Goura* research are coming from the captivity or zoos. For instance, Rotterdam Zoo in Netherland always releases the important European Studbook of *Goura,* because they have been doing the long term research on Genus *Goura* (Nijboer and Damen 2000, Beltermann and Pott 2008). Several research on *Goura* species which have been done in Papua include study on feeding behaviour, breeding behaviour and propagation of *Goura victoria* (Setio *et al* 1996, Roembino 1997), food palatability of *Goura cristata and Goura victoria* (Tribisono 2002, Notanubun 2002), *Goura cristata* genetics (Kilmaskossu 2001), and genetics of all species of

Genus *Goura* (Siahaan 2006), and study on birds hunting including *Goura victoria* (Sada 2005, Mahuse 2006, Pangau-Adam and Noske 2010).

CHAPTER 3: RESEARCH METHODS

3.1. Study area

3.1.1. Papua Province

Recently, Papua is administratively divided into Province of Papua and Province of West Papua (figure 3.2: the white area is the West Papua Province and the other color is Papua Province). Papua Province lies between 2°25'-9°S and 130°-141°E, with total land cover of approximately 317.062 km^2 or 17.04% of Indonesia area. Since 2009, Papua Province is divided into 28 Regencies and one Municipality.

Papua has tropical climate with two seasons a year, dry season that last from June to September and rainy season from December to April. Daily temperature varies around 14.8 – 27.5°C at night and 26 - 32°C during the day (BPS 2010[b]). The average rainfall of Papua province fluctuates between 1381 mm- 4014 mm annually and rainy days can reach 160 – 281 days a year (BPS 2010[b]).

Figure 3.1 Map of Papua Province (Source: http://papua.bps.go.id).

Papua region is still largely forested with a variety purpose of forest utilization both by the society and the Government. According to the latest data from Balai Pemantapan Kawasan Hutan Wilayah X Jayapura (*Forest Observation Agency Area X Jayapura*) in 2010, the total forest area of Papua province is

31.773.063 ha. This area consists of 12.639.840 ha or about 39.78% declared as production forest, 1.769.221 ha (5.57%) limited production forests, 6.440.282 ha (20.27%) exchangeable production forests, and 13.906.393 ha (43.77%) as protection forest and conservation area. The rest area or a total of 1.512.690 ha (4.76%) is classified as other designated area including lands and water. The same source also classified the land cover of Papua into several areas include 20.971.610 ha (66% of the whole land cover) primary forest 3.442.842 ha (10.83%) secondary forest, 5.308.693 ha (16.71%) non-forest area, 1.475.230 ha (4.64%) unidentified cloud covered area and 574.688 (1.81%) ha water area, (appendix table 6).

3.1.2 Research location

The study was concentrated in the northern part of the Papua Province, in four different regencies, Mamberamo Raya, Jayapura, Sarmi, and Supiori, where the presence of Victoria Crowned Pigeon was recorded (Beehler *et al* 1986). Detailed observations on the population density of *Goura victoria* was conducted in four forest areas, Buare forest in Mamberamo Raya, Northern Supiori forest in Supiori, Unurumguay forest in Jayapura and Bonggo forest in Sarmi regency (figure 3.3).

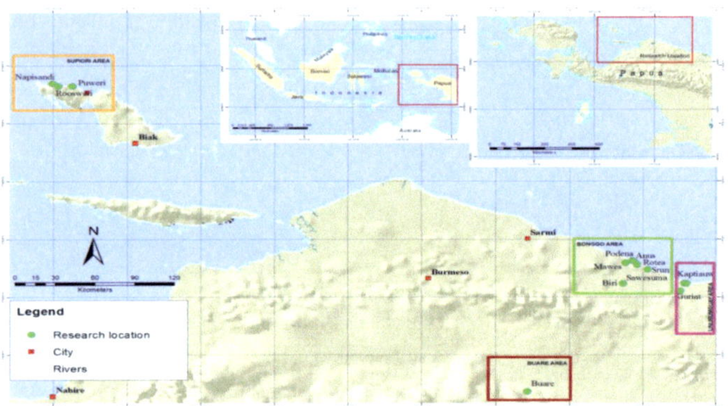

Figure 3.2 Map of the northern part of Papua Province and the location of field studies

3.1.2.1 District of Mamberamo Tengah - Mamberamo Raya Regency

The Regency of Mamberamo Raya is new regency in Papua Province, which is expanded from Sarmi Regency. The partition was based on the Legal Acts No. 19/2007. Mamberamo Raya Regency consists of 8 district and 59 villages (http://www. mamberamorayakab.go.id/). This regency lies between $01°28'$ – $03°50'$ S and $137°46'$ – $140°19'$ E.

Buare village as the study site is located at $3°18'56.8"$S and $138°42'51.0"$E. It is one of the remote villages and a part of District Mamberamo Tengah, Mamberamo Raya Regency. The village can only be reached from Dabra (District Capital) by traditional boat or 'ketinting' up to the Mamberamo River and turn to the Buare River, which is taken approximately 3 – 5 hours trip. In the rainy season, Buare's villagers usually have to walk through the forest to get to their village. Dabra can be reached from Jayapura (the capital city of Papua province) by flying with light aircraft or small propeller aircraft, or by long time-boarding in small-ships from the mouth of Mamberamo River.

Buare is only a small settlement inhabited by 15 families with less than 70 people. As stated before, this village was an expansion from Dabra, so it has no available supporting facilities. Houses at Buare village are very simple, built on stilts, with floors made from barks of Nibung palm (*Oncosperma tigillaria*), roofs from woven grasses or 'alang-alang' (*Imperata cylindrica*), and most of the house are built without walls. The settlement was only temporary until the infrastructures were built up by the Government. Consequently, the entire population of this village lived temporarily in Dabra. Commonly, Buare villagers only stay in their village for about 3-4 months a year. The main livelihoods of the villagers include the practice of subsistence hunting and non-intensive agriculture. The wild meat that obtained from hunting is mostly used for self-consumption, but sometimes it is also sold on the market day in Dabra, if they need fresh money.

The whole area of Mamberamo Raya Regency lies in the Mamberamo watershed. The Mamberamo-Foja Wildlife Sanctuary was established in this watershed and was declared under the Decree of Republic Indonesia Minister of

Agriculture No. 820/1982 (http://bksdapapua.net/index.php/suakamarga satwamamberamofoja.html).

In general, there is low level of human interference in Buare's forest, so the large area of primary forest still remains. The forest is located far from the village and still difficult to be accessed, so commonly only the local people or tribal land-owners can enter the forest. The forests have a high potential of wood, for example the commercial-valuable ironwood (*Intsia bijuga* L.). Also, there are many wild animals inhabit the forest, for instance Victoria Crowned Pigeon, wild hog, tree kangaroo, birds of paradise, Blyth's hornbill, and northern cassowary. (http://www. mamberamorayakab.go.id/).

The study site in Buare was established in Buare watersheds. In this site, there were small-scale traditional farms where the local people usually plant and harvest their crops every three or six months per year. This condition may occur because the Buare tribes stay in their village less than three months.

3.1.2.2. District of Unurumguay-Jayapura Regency

The study sites in District Unurumguay, Jayapura Resgency were located in Guriath village and Sawesuma village. Each village has different features.

The village of Guriath lies on 02°26'39.2"S and 139°45'30.3"E. The population is about 300 people or approximately 60 families, and most of the villagers prefer to stay near to the main road. The main livelihoods are hunting and subsistence gathering-farming. In order to earn money, they generally collect pebbles and stones from Tuarim river (under Ondoafi's or village chief's permission), then sell these materials to settlement developers. Some of them are working as daily labors in road construction companies. This village does not have sufficient infrastructure facilities like village office. Consequently, most administrations and activities of the community were centered on the house of the Village's Secretary.

The next study site, Sawesuma village is located about ten kilometers from Guriath village, and lies on 02°22'51.4"S and 139°47'33.5"E. These two villages are separated by the Trans Irian highway that connecting Jayapura city, Jayapura Regency and Sarmi Regency. Unlike Guriath, Sawesuma village has already a complete infrastructure. There are 95 families or approximately 400 people living in Sawesuma village (Jayapura Regency in figures 2010, http://jayapurakab. bps.go.id).

The livelihood of Sawesuma villagers consists of hunting, gathering-farming and dredging sago starch. To earn money, the villagers usually work as wages labors on the nearest construction company and also sell stones and pebbles. Observation showed that villagers from both villages prefer to work as labors, because it is easier and faster to achieve some money compared to other jobs. Local people in both villages are planting cocoa in their farm as supported by the farming program from government of Jayapura Regency. These plantations are usually located in the edge of village or even in the forest area. Commonly, economic activities occur among the villagers themselves, or only with neighbor villages due to the absence of market. For daily purposes, the villagers rely on mobiles vendors or kiosks run by the people from other Indonesian islands which sell various daily needs with high price. The distance between villages to district center or regency capital, high cost of local transportation, and lack of public transport vehicles are the main problems on the accessibility for the villagers to other places (Village secretary of Guriath and Sawesuma, *pers.comm.*)

The forest area around both villages can be classified as secondary forest. This forest had been a logging concession area of PT Wapoga Mutiara Timber since 1980's until 20 years ago. Besides, all forest areas are commonly used by local people for shifting cultivation, gathering forest products and hunting. The forest is a natural habitat for several Papuan wildlife including Victoria crowned pigeon, paradise bird, two species of Megapodes, northern cassowary, fruit pigeons, wallaby, kangaroo and wild hog. Due to the riches of wildlife, this forest area is often visited by outsiders for hunting wild hog or deer, with the permission from village's chief or Ondoafi.

The district of Unurumguay was basically a logging concession forest of PT Wapoga Mutiara Timber as well, but since 20 years ago the logging activity has already ended. Recently, this forest area is frequently used by local people for their daily activities like hunting, traditional farming, and also for logging activities run privately by the chef of tribe (Ondoafi). Many areas so-called 'kebun' or small farms run by the villagers can be found in all forest area. In addition, small-scale illegal logging by the local people and the transmigration people from Java Island has emerged in this area and now getting increased. Due to development in Papua, a highway called 'Trans Papua' was already built passing through the forest area, to link Jayapura Regency and Sarmi Regency. A number of new resettlements in these regencies can be found along this main road.

3.1.2.3. District of Bonggo, Sarmi Regency

Sarmi Regency is anew regency that was expanded from Jayapura Regency according to the Indonesian Law No. 26/2002. The regency is located at $138°05'$-$140°30'E$ and $1°35'$- $3°35'S$ and consists of 8 districts and 84 villages. It is bordered by Pacific Ocean in the north, Mamberamo Raya Regency and Tolikara Regency in the south, and Mamberamo Raya Regency and Jayapura Regency in the east and west. The area's average temperature is around $21.9°C$, average humidity is 85.3%, and 145 mm of the average rainfall (http://www.sarmikab.go.id). Bonggo District has 12 villages, and seven of the villages were selected as the study sites (table 3.1).

Table 3.1 The position of the seven villages as study sites in
District Bonggo-Sarmi Regency

Villages	Geographic Position
Kaptiauw	$02°22'51.4"S$ and $139°47'58.68"E$
Srum	$02°15'33.32"S$ and $139°32'00.11"E$
Anus	$02°11'13.01"S$ and $139°26'19.78"E$
Rotea	$02°12'56.76"S$ and $139°27'40.64"E$
Biri	$02°22'50.63"S$ and $139°21'52.45"E$
Mawes	$02°12'02.96"S$ and $139°22'55.08"E$
Podena	$02°10'56.67"S$ and $139°25'44.68"E$

The construction of the Trans Papua highway between Jayapura city, Jayapura Regency and Sarmi Regency has a distinctive impact. Due to the presence of the Highways, the government of Jayapura Regency has established the policies to declare Bonggo as transmigration areas. Then after the subdivision of regency and regency development, government of Sarmi Regency decided to resettle the people from several villages in the remote areas and the coasts to the areas along the Trans Papua highway. This decision was called local transmigration program. Consequently, local people began to recognize the small-scale farming practices like planting crops, rising livestock and also trade system.

Furthermore, the government has developed a program of cocoa plantation, with the aim to raise local community welfare, as well as to reduce their activities on wildlife hunting, collecting non forest timber products and shifting cultivation in the forest. Unfortunately, the lack of transportation, the absent of market near villages, and expensive prices of basic commodities, remain the major obstacles in such development efforts. These conditions were contrast if compared to the area's valuable diversity. Mamberamo Raya Regency (including Buare village), Sarmi Regency (including Bonggo District) and Jayapura regency (including Unurumguay District) are located in Endemic Birds Area (EBA) of lowland tropical rain forest in the northern part of Papua. This area holds the highest potential of nine species of birds in Restricted Range Areas (RRA) category (figure 3.4.). *Goura victoria*, cassowary birds and other endemic birds can be encountered in this area (Beehler *et al* 1986).

Figure 3.3 Endemic Bird Area in lowland forest of northern Papua-
Indonesia
(Source:http://burung.org/Daerah-Burung_Endemik/176-dataran-rendah-dibagian-utara-papua.html).

Additionally, Regency of Jayapura and Sarmi are located in the north coast, where the human population is high. Therefore, the lowland forest is threatened through the large-scale of human interferences and deforestation. These activities including opening of the new territories for new districts, new villages, building the infrastructures, resettlements of the local residents, establishing oil palm plantation and logging concessions. It can also be predicted that development of regencies such as Mamberamo Raya and some other new regencies within watershed area of Mamberamo River can lead to the increasing of forest degradation. The establishment of new regencies may facilitate the access to remote and isolated forest areas, and consequently forest exploitation will be expanded to the valuable forest ecosystem which is need to be protected.

The forest in Bonggo areas is a secondary forest that located in Bonggo Mountains. This forest was logged under logging concession (HPH) of PT. Wapoga Mutiara Timber, since 1980's. In 2010 the logging license for this company was terminated, but currently this logging company is await for a new license to operate again.

3.1.2.4. District Supiori Utara-Supiori Regency

Similar with the Regency of Mamberamo Raya and Sarmi, Supiori regency was also expanded from Biak-Numfor Regency according to Indonesian Act No. 35 /2003, and currently has 5 districts and 38 villages (http://pemdasupiori-papua.com/). Three villages in Supiori regency were decided as the study sites, Puweri (lies on 0°40'31.56"S and 135°37'51.43"E), Rosweri (0°40'30.71"S and 135°31'58.85"E) and Napisndi (0°40'30.71"S and 135°31'58.85"E).

The regency of Supiori consists of Supiori Island and Mapia Island. Most area of Supiori Island or about 42.000 ha (95% of land area) is a nature reserve, declared by Minister of Forestry Decree No. 26 /1988 (http://gispapua.com/ accessed on June 27, 2012). The primary forest of the island is mainly concentrated in the northern Supiori. This forest is still a part of nature reserve (Saaroni and Simbolon 1998), and is dominated by several plant species like the ironwood (*Intsia* spp), *Pometia* spp, damar wood (*Agathis labillaldieri*), chinawood (group of *Dacrydium* spp, *Podocarpus* spp and *Phyllocladus* spp) and

Ketapang (*Terminalia catappa*). The forest area has high diversity of bird species, with at least 13 species of bird habitats are categorized as Restricted Range Area (EBA,http://burung.org/daerah-burung-Endemik/174-Biak-Numfor.html/ accessed on June 27, 2012 *see* figure 3.5). The presence of Victoria crowned pigeon is also predicted in the Island, although it might be introduced from Yapen Island (Rand and Gilliard 1967). Some wild animals such as northern cassowary, wild hog, phalanger and *Goura victoria* have become the target species for wildlife hunting.

Figure 3.4 Endemic Bird Area in the lowlands of northern Papua-
Indonesia
(Source:http://burung.org/Daerah-Burung_Endemik/174-Biak- Numfor.html)

Commonly, the Supiori residents set up their farming fields in the mainland and in nature reserve area through practicing slash and burn cultivation system. They also usually pursue hunting activities and shifting cultivation. These activities occurred more frequent in the fisherman villages, because they temporally change their works during the bad season. Wildlife hunting and farming are basically run by the villagers for self-consumption. However, these practices have become the main causes of forest degradation in Supiori areas (EBA   http://burung.org/daerah-burung-Endemik/174-Biak-Numfor.   html/   - accessed June 27, 2012; Saaroni and Simbolon, 1998). Furthermore, similar with other areas in Papua, Supiori forest will be degraded rapidly due to the regional population growth (Hope 2007).

Northern Supiori area is a part of Bon Supiori Nature Reserve. This region has a variety of ecosystem, from coast to the mountains with an altitude of more than 800 meters above sea level, and about 40 - 65% slope or more, specifically at the south of Bon Supiori (http://www.papua.go.id/view-detail-peta-7/lereng.html accessed 19 September 2013). Four districts and all villages are situated on the coastal line, near to the foothills of Supiori Mountain. Easy access to the forest might support forest utilization and encroachment from the villagers.

3.2. Data collection and analysis

3.2.1. Time table of data collecting

The field work was carried out in four different periods; (1) August to November 2007 in Northern Supiori area, (2) March to April 2008 in Buare-Mamberamo areas, (3) August to November 2009 in Unurumguay area and (4) August to November 2010 in Bonggo area.

3.2.2. Habitat parameters

Habitat parameters were assessed in order to know the composition and vegetation structure of forest in each study site, and to describe the ecological and habitat quality of the site.

In each study site, as many as 25 randomly long-line transects for vegetation analysis of 20 x 100 meters with 20 m x 20 m plots were established. Trees with a diameter at breast high > 10 cm and more than one meter height within each plot were measured and identified. Furthermore, floristic structure was assessed quantitatively by calculating the Important Value Index (IVI) for each species in each study site (Dumbois and Ellenberg 1994, Brower *et al* 1997). The IVI represents the sum of the value of Relative density (RD), Relative frequency (RF) and Relative Dominance (RDo), which are determined by the following equation:

$$\text{Relative density} = \frac{Number\ of\ individuals\ of\ a\ taxon}{Total\ number\ of\ individuals} \times 100$$

$$\text{Relative frequency} = \frac{Number\ of\ plots\ containing\ a\ taxon}{Total\ frequencies\ of\ all\ taxa} \times 100$$

$$\text{Relative density } = \frac{Basal\ area\ of\ a\ taxon}{Total\ basal\ area\ of\ taxa} \times 100$$

Forest structure of each study site was assessed by comparing the distribution of canopy heights and trunk diameter classes. Further analysis of height distribution and diameter distribution was followed Hadi *et al* (2009). In addition, Shannon-Wiener index (H') and its variance of H' (Magurran 1998) were calculated to determine the plant diversity between the study sites. Tree species were identified based on the information from taxonomist of Forest Research and Development Institute Manokwari. For the species that were unable to identify in the field, the voucher specimens were collected and sent to the Herbarium of Bogoriense, Bogor, Indonesia.

3.2.3. Population of *Goura victoria*

Goura surveys were carried out at four sites using line transect methods (Buckland *et al*, 2001) and 45 transects were set aside in all study sites. The researcher walked along the transect line and recorded the perpendicular distance between detection points and transect line. Surveys were done four days per week on 06.00 - 16.00 by the field team.

Population density (D) of *G.victoria* was calculated using Line Transects Formula (Buckland *et al* 2009) as follows:

$$\textbf{D = (N x 1000)/2LW}$$

Where the width of the transect line W (in meters) is two times of the mean distance of transect line. L is line length in kilometers, and N is the number of bird seen or heard. The Distance 5.0 Release 2 program was used for data analyzing (Laake *et al* 2006, CREEM 2009).

3.2.4. Hunting activities by local people

In order to determine the activities of local hunter, used semi structural interviews with questionnaires. The study sites or villages were chosen purposively due to the most accessible and feasible reason. All sites should also secure enough to be studied according to the availability of time and resources. All of the villages selected as study sites were situated around the forest area.

Selections of villages and hunters as respondents were taken based on the information from key informants (villages headmen, head of tribe or 'Ondoafi' and church elderly). The levels of hunting activity by local people were fluctuated seasonally therefore the interviews were conducted after the completion of survey on *G.victoria*.

The interviews were conducted on 151 respondents who live in 13 villages of four districts in four regencies. The location of those villages were mapped (figure 3.2), and the information of each village was described in table 3.2.

Table 3.2 The villages, individual surveyed, indigenous communities surveyed and belief system where the surveys were undertaken.

No	Regencies	District	Villages Surveyed	Village name	Individual Surveyed	Indigenous communities	Belief System
1.	Mamberamo Raya	Mamberamo Tengah	1	Buare	8	Buare	Christianity
2.	Jayapura	Unurumguay	2	Guriath	17	Unurumguay	Christianity
				Sawesuma	16		
3.	Sarmi	Bonggo	7	Kaptiauw	11	Bonggo	Christianity
				Srum	11		
				Anus	12		
				Rotea	9		
				Biri	13		
				Mawes	13		
				Podena	9		
4.	Supiori	Supiori Utara	3	Puweri	15	Biak	Christianity
				Rusweri	10		
				Napisndi	7		

In each community or village, hunter (one man hunter per household) was interviewed. The questionnaire was focused on the following questions:

a. Frequency of hunting (categorized as : often, sometimes or rare)
b. Number of person participating (single or groups)
c. Hunting distance (close : < 2 km, middle: 3-5 km, far: >5 km)
d. Frequency of bird meat consumed (commonly, occasionally and rarely)
e. Hunting time (day or night)
f. Hunting season
g. Other species hunted

h. Hunting methods (traps, gun, dogs)

i. Numbers of *G.victoria* captured

j. Hunting purposes (money, meat or both)

k. Bird age and sex of *G.victoria*

l. Utilization of bird parts (all part or only carcasses)

m. Permission, protection and ban of hunting on *G.victoria*

n. The age of hunter

Information on all hunted animals was collected to gain insight into hunting patterns and determined the relative importance of *G.victoria*. Respondents were also asked about forthcoming hunting trips or orders from traders. In addition to *G.victoria*, all captured or killed animals were later identified, and if brought back to the community (with the hunter's permission) their weight, actual price and selling price were noted as well.

Non parametric statistics include Mann-Whitney U test, Kruskal-Wallis test and Multiple Linier Regression followed Aiyadurai *et al* (2010) and O'Brien *et al* (1998) were used to assess several variables in hunting. All analysis was run with SPSS Release 17.

3.2.5. Estimation on the sustainable harvest of *Goura victoria*

The r-values for population growth were calculated in order to estimate the sustainable harvest rates of *G.victoria*. The r-value is the maximum intrinsic rate of population increase of a population that is not limited by food, space, resource competition, or predation. This value was calculated according to the following formula (Bodmer *et al* 1997, Robinson 2000):

$$1 = e^{-rmax} + be^{-rmax(a)} - be^{-rmax(w+1)}$$

Where: e = 2.7128, a = the species-specific age at first reproduction, w = the age of last reproduction and b = the annual birth rate of female offspring. The value of a, w and b of *G.victoria* was obtained from studies on *Goura* in captivity (Belterman and Poot 2008, Gibbs *et al* 2001, Baptista *et al* 1997, Coates, 1985).

Afterwards, the finite rate of increase (λ) was calculated to estimate the population growth over time. This value was converted from the instantaneous rate of increase (λ_{max}) according to formula from Noss (2000) and Robinson (2000):

$$\lambda_{max} = \lambda, \text{ so as: } \lambda = e^r$$

The λ-value is to estimate the maximum annual production at the observed density $(P_{max(D)})$, using formula from Robinson and Bennett (2000[b]) and Robinson (2000)

$$P_{max(D)} = [(0.6D \times \lambda_{max}) - 0.6D]$$

$$= (\lambda_{max} - 1)(0.6D)$$

Where **D** is represents the observed value from the population density of *Goura* in this study.

Specifically for *G.victoria*, the factor f_{RR} is defined as **0.2** representing the production in long-lived species, for those whose age of last reproduction is over 10 years. Finally, all the earlier values were used to estimation the maximum sustainable harvest (based on Begazo and Bodmer 1998) follows:

Maximum sustainable harvest = (P_{maxD}) (0.2)

To assess hunting sustainability in each study site, the estimation of maximum sustainable annual harvest was compared with the value of maximum current annual harvest. Hunting practice on *Goura* is unsustainable if the value of maximum current annual harvest exceeds the value of maximum sustainable annual harvest.

All data analysis was run with Excel program.

CHAPTER 4: RESULT

4.1. Floristic and structural composition of the forest in each study site

4.1.1. Floristic composition

Vegetation analysis showed the different results in four sites. A total of 58 tree species in 38 families was found in Buare and 57 tree species (38 families) in Supiori. These results were slightly different from those recorded in Unurumguay, (39 tree species in 25 families) and in Bonggo (34 tree species in 22 families). The total number of tree species found in the whole study sites was 188 tree species in 123 families (appendix 1, 2, 3 and 4). The percentage of tree species recorded from each study site is showed in figure 4.1.

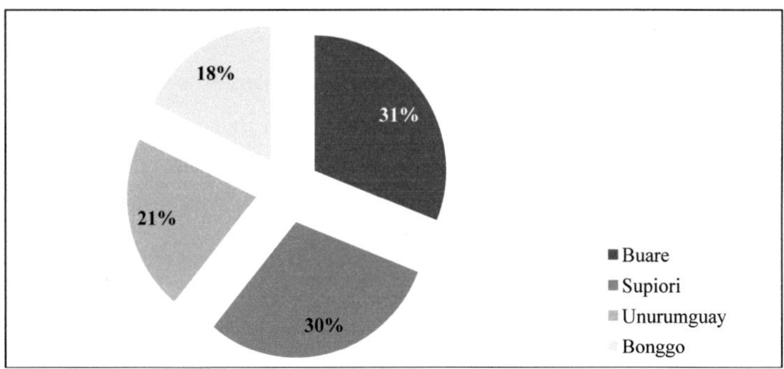

Figure 4.1 Percentage of tree species encountered in each study site.

Calculation of the value of density, frequency and dominance level from each tree species showed a slightly different result on the Important Value Index for each study site (table 4.1).

Table 4.1 The value of Relative Density (RD), Relative Frequency (RF), Relative Dominance (RDo), and Important Value Index (IVI) in each study site.

Variables	Study site			
	Buare	**Supiori**	**Unurumguay**	**Bonggo**
RD (%)	100	100	100	100
RF (%)	100	99	100	100
RDo (%)	100	100	100	100
IVI (%)	300	299	300	300

Table 4.1 showed that the value of RD from each study site reached the maximum value of 100%, similar with the RDo value. The RF value is almost the same, which ranged between 99% and 100% and the Important Value Index (IVI) ranged around 300%.

Composition of tree species in each study site also varied as showed through the Shannon-Wiener Diversity Index; H'= 3.55 in Buare forest, 3.45 in Supiori, 3.09 in Unurumguay and 3,00 in Bonggo. Although the diversity in Buare seems more diverse than that in other sites, it is statistically not significant, because the values of H' of all study sites are in the range between 1– 4.5 (Magurran 1987).

The t-test was used to know the variation of diversity index between study sites (Magurran 1987). There is no significant difference of diversity index between Buare and Supiori forest, t = 1.09 (P >0.01), and between Unurumguay and Bonggo forest, t =0.37 (P >0.01). The comparison between the other sites are significantly different; between Buare and Bonggo, t= 5.36 (P<0.01), Buare and Unurumguay t= 4.49 (P<0.01), Supiori and Bonggo t= 4.74 (P<0.01) and between Supiori and Unurumguay t= 3.79 (P< 0.01) (figure 4.2). These results suggested that the plant diversity of Buare, Supiori and Unurumguay forests area higher than that in Bonggo forest (Fachrul 2008).

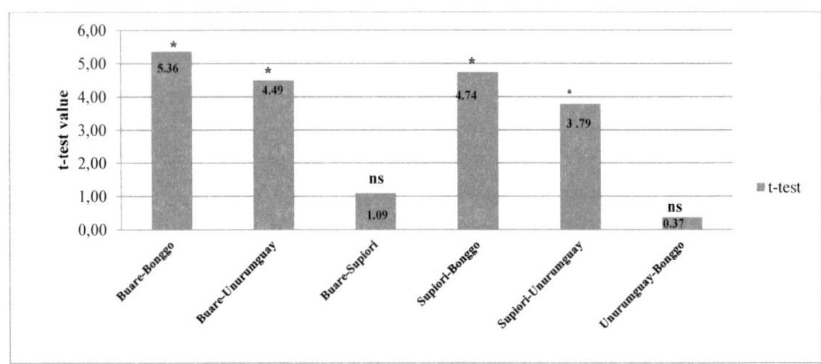

Figure 4.2 Variance of plant Diversity Index between study sites (based on t-test, *= significant, ns= non significant).

The seven most dominant tree species based on the Important Value Index were varied between study sites (table 4.2). These species belong to different families, with Euphorbiaceae family as the most common family encountered in all study sites. The table 4.2 showed that *Pimeliodendron amboinicum* Hassk was dominant tree species in forest area of Buare, Supiori and Unurumguay, while *Pometia* spp. (*Pometia pinnata* and *Pometia* sp.) dominated forest area in Buare, Unurumguay and Bonggo.

Table 4.2 The seven most dominant tree species encountered in each study site.

Study Sites	Tree Species	Family	RD (%)	RF (%)	RDo(%)	IVI (%)
Buare	*Pimeliodendron amboinicum* Hassk	Euphorbiaceae	11.67	11.49	15.37	38.53
	Eugenia anomalia Lauth	Myrtaceae	6.67	6.32	12.65	25.64
	Syzygium sp	Myrtaceae	7.22	6.90	9.44	23.56
	Intsia spp	Fabaceae	6.67	6.90	6.00	19.56
	Pometia pinnata	Sapindaceae	5.56	5.75	4.57	15.87
	Myristica papuana	Myristicaceae	6.15	6.36	2.10	14.61
	Terminalia microcarpa Decne	Combretaceae	2.78	2.30	4.65	9.73
Supiori	*Canarium indicum* L.	Burseraceae	5.82	6.24	12.52	24.58
	Eugenia anomala Lauth	Myrtaceae	8.73	6.71	8.98	24.42
	Planchonella anteridifera	Sapotaceae	5.82	4.80	11.94	22.55
	Homonoia javanensis M.A.	Euphorbiaceae	10.91	6.71	1.62	19.24
	Callophyllum inophyllum Linn	Calophyllaceae	6.91	6.71	5.24	18.86
	Pimeliodendron amboinicum Hassk	Euphorbiaceae	6.18	5.76	5.35	17.29
	Blumeodendron sp	Euphorbiaceae	5.45	3.84	2.37	11.66
Unurumguay	*Pimeliodendron amboinicum* Hassk	Euphorbiaceae	11.30	11.31	17.98	40.59
	Pometia spp	Fabaceaec	12.99	13.69	12.09	38.78
	Myristica papuana Kunth	Myristicaceae	10.17	10.71	6.05	26.94
	Pterygotha horsfildea	Sterculiaceae	8.47	7;74	7.62	23.83
	Intsia spp	Fabaceae	5.65	5.95	6.38	17.98
	Dracontomelum edule Merr	Anacardiaceae	3.95	4.17	9.61	17.73
	Campnosperma auriculatum	Anacardiaceae	6.21	6.55	4.21	16.97
Bonggo	*Syzygium* spp	Myrtaceae	13.16	13.16	8.93	35.25
	Myristica papuana	Myristicaceae	11.84	11.84	10.11	33.79
	Pometia spp	Fabaceae	10.53	10.53	9.32	30.38
	Blumeodendron sp	Euphorbiaceae	7.24	7.24	7.48	21.96
	Palaquium amboinensis	Sapotaceae	7.24	7.24	6.31	20.78
	Campnosperma auriculatum (Bl) Hook.f.	Anacardiaceae	5.92	5.92	7.92	19.76
	Pterocarpus indicus Willd	Fabaceae	5.92	5.92	6.16	18.00

The whole tree species encountered in all study sites could also be classified based on their function as bird food sources. About 51 tree species were the food plants of frugivorous birds from eight families including Columbidae that inhabit in the Australasia region (Snow 1981). A total of 19 tree species in 12 families in Buare, 21 species in 12 families in Supiori, 19 species in 10 families in Unurumguay and 12 species in 11 families in Bonggo could be classified as bird food trees. There were a total of 51 species in 14 families considered as food source trees found in all study sites. These numbers were about 35.92% of 142 species and 13.73% of 102 families which have been listed as bird food source for Australasia region (Snow 1981).

4.1.2. Forest structure composition

The measurement of diameter at breast height (dbh) and tree height class distribution were used to describe structural composition of forest area in each study site. Figure 4.3 shows that in four study sites, approximately 80% of forest vegetation was represented by individual trees with dbh-value less than 30 cm. The individual trees in Bonggo forest were dominated by the trees in diameter class of 15-19 cm, 24-29 cm, and 25-29 cm, but no tree in diameter class more than 35 cm. The trees in other study sites were found in all diameter classes including diameter more than 35 cm, but the number of individual trees were decreasing from the lower diameter to the higher diameter values.

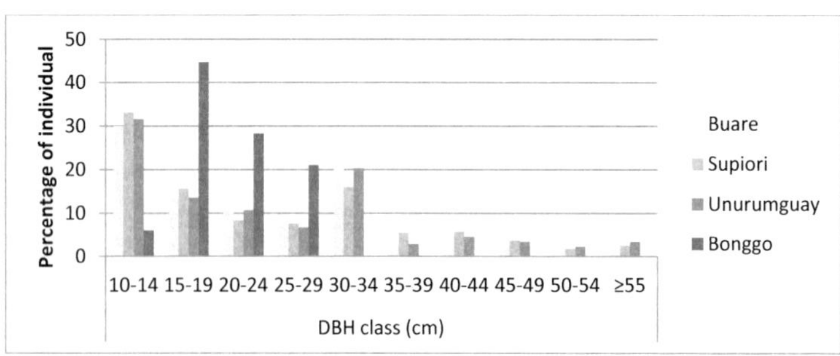

Figure 4.3 Distribution of trees in diameter at breast height (dbh) classes in each study site.

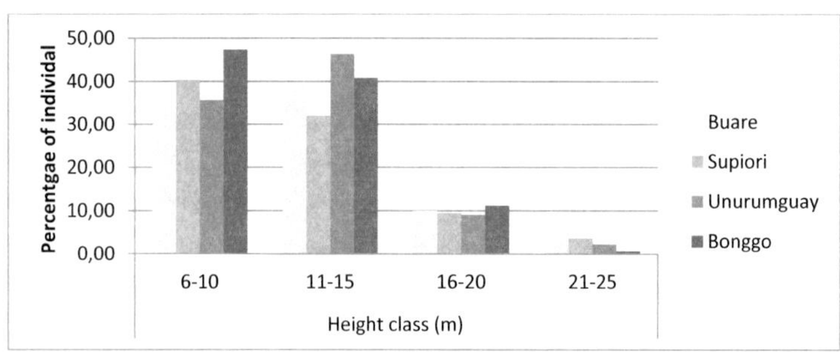

Figure 4.4 Distribution of trees in height classes in each study site.

The height class distribution of trees in all study sites can be classified in some groups (figure 4.4), most of the trees (78.81%) belonged to the height class of 6-10 m and 11-15 m, while the rest (12.39%) distributed into the height class of 16- 20 m and 21 – 25 m. In other words, the height distribution shows that forest in all study sites has similar structure that ranged from height classes of 6-10 m, 11-15 m, and 16-20 m, and the number of trees decreases along with the increase of height class. The distribution of height class shows that vegetation in all study sites can be classified into *mid-lower canopy* class (Richards 1996).

4.2. Estimation of density population of *Goura victoria*

The populations of *G.victoria* were observed in a total of 45 transects, 135 km of transect length and 232 observation effort. The value of bird density for each site was calculated using Multi Covariate Distance Sampling (MCDS) with Distance Program Release 5.0 (Laake *et al* 2006). Density estimates of the bird were different among study sites, 41.74 birds.km^{-2} in Buare, 40.29 birds.km^{-2} in Supiori, 30.80 birds.km^{-2} in Unurumguay, and 13.10 birds.km^{-2} in Bonggo.

Density estimates of *G.victoria* in Buare, Supiori and Unurumguay were not significantly different (Z test< P 0.05) but those values were significantly different from the density estimates of *G.victoria* in Bonggo area (figure 4.5).

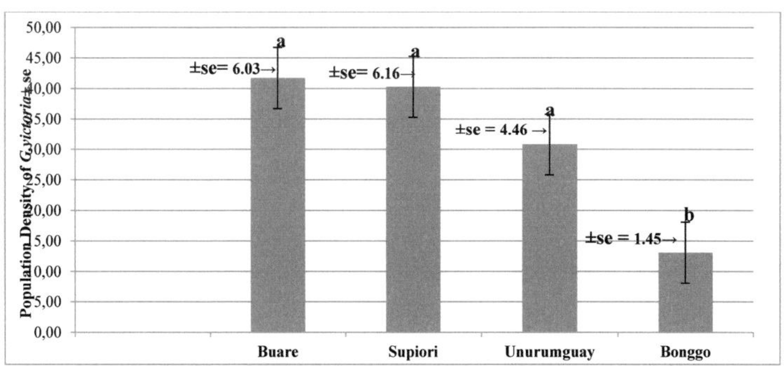

Figure 4.5 Population density estimation of *G.victoria* in each study site
        (± se: standard error).

Figure 4.6 illustrated the result of variance of population density of *G.victoria* based on Z-test (95% confidential interval-CI) that was carried out to determine the detail variation of population density of *G.victoria* between each site.

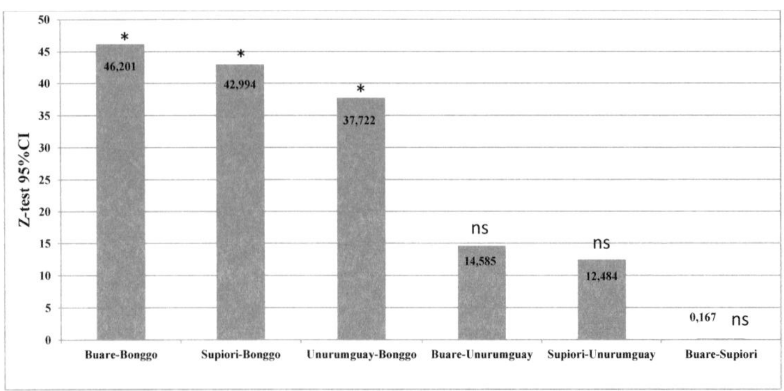

Figure 4.6 Variance of Population Density of *G.victoria* between study sites
(* = significant, ns= non significant).

Furthermore, the estimation of *G.victoria* population was calculated based on the size of hunting area in each study site (table 4.3). The formula for estimating size of hunting area is πR^2, where R is the maximum hunting distance in every study site. The results showed that the population of *Goura victoria* in each hunting area varied from 2451 to 4380 birds in Buare site, 3344 to 6206 birds in Supiori site, in Unurumguay site was 2589 to 4682 birds while in Bonggo site was 3266 to 5178 birds.

Table 4.3 The estimation of population density of *G.victoria* in each study site.

Study site and Regency	Density estimate (bird/km²)	Hunting area size for each study site (km²)	Population estimate (birds, 95% CI)
Buare (Mamberamo Raya)	41.74	78.5	2451-4380
Supiori (Supiori)	40.29	113.04	3344 – 6206
Unurumguay (Jayapura)	30.80	113.04	2589 – 4682
Bonggo (Sarmi)	13.10	314	3266 - 5178

Notes: 95% CI = 95% Confidential Interval

Plant diversity (Shannon-Wiener Indexes) was strongly positively correlated with population density of *G.victoria* as indicated by the Pearson correlation coefficient, r^2 = 0.73 (P<0.05). This may show that 73% of the presence of *G.victoria* was affected by the vegetation structure and 27% by other factors. The other factors may include activities of local people living in surroundings of the study sites.

4.3. Hunting activity of Papuan hunter

In order to obtain comprehensive information on hunting activity, specifically on *G.victoria*, the interviews were directed not only to the hunters as the main respondents, but also to other respondents including Ondoafi (head of tribes), heads of the villages and village elders. Interview was also done with head of the districts for collecting data on governance, licensing, other important information related to hunting activities, taboos, and daily activities of hunters. The result of interviews is showed in the table 4.4.

Table 4.4 The characteristics of hunters and hunting system in each study site

Parameters		Study site/Forest area			
		Buare	Supiori	Unurumguay	Bonggo
The numbers of respondent (n)		8	32	33	78
Age of hunter (years old)		25 - 50	27 – 59	29 – 60	27 - 60
Frequency of hunting :	often (twice a week)	38%	44%	36%	68%
	frequently (once a weeks)	63%	38%	42%	15%
	rare (once a month)	0%	21%	21%	17%
Size of hunting group:	single person	63%	81%	88%	77%
	group ($\geq$2)	38%	19%	12%	23%
Successful on hunting *Goura:*	successful	62.50%	62.50%	48.48%	52.56%
	Not successful	37.50%	37.50%	51.52%	47.44%
Frequency of *Goura* meat consumption:					
	Commonly (once a week)	0%	0%	0%	0%
	Occasionaly(once a month)	38%	31%	24%	21%
	Rarely (once per three month)	63%	69%	76%	79%
Quantity of capturing *Goura*:	none (0 individuals)	38%	38%	52%	47%
	one (1 individuals)	63%	50%	48%	46%
	two or more ($\geq$2 individuals)	0%	13%	0%	7%
Using air gun	yes	0%	20.88%	3.03%	8.97%
	no	100%	79.12%	96.97%	91.03%
Using dogs:	yes	12.50%	12.50%	12.12%	11.54%
	no	87.50%	87.50%	87.88%	88.46%
Distance of hunting:	Far	0%	49.99%	21.21%	25.64%
	Middle	62.50%	21.88%	60.61%	30.77%
	Near	37.50%	28.13%	18.18%	43.59%
Preferred time for hunting:	day	0%	15.38%	0%	34.38%
	night	100%	84.62%	100%	65.65%
The attitude of protecting *Goura*:	agree	100%	40.63%	69.70%	62.82%
	Not agree	0%	59.38%	30.30%	37.18%
Difficulties when hunting *Goura*:	yes	0%	15.63%	45.45%	32.05%
	no	100%	84.38%	54.55%	67.95%
Using foot snares	yes	100%	100%	100%	100%
Preferred season for hunting	yes	100%	100%	100%	100%
Selected age of *Goura*	no	100%	100%	100%	100%
Selected sex of *Goura*	no	100%	100%	100%	100%
Selected *Goura* meat	yes	100%	100%	100%	100%
Purposes of hunting on *Goura*: cash money and meat (yes)		100%	100%	100%	100%

Notes: The structure of the table followed O'Brien *et al* (1998).

4.3.1. Status of hunters

There were 151 hunters who participated in the interviews about hunting system in all study sites. A total of 143 respondents (94.71%) were working as farmer and hunter, and only eight people (5.29%) were government employee and also practicing wildlife hunting.

The hunters usually started their hunting practices at their early years, about 11 years old, and practicing wildlife hunting until they are in the old age. Based on the interview, the average age of hunters was 44.50 (±8.40) years old. In all study sites, the range of hunter ages interviewed was between 25 and 60 years old. The largest number of respondents or hunters was in the class of 45 – 49 years old (23.18%) and the smallest number (1.32%) of hunters found in the age class of 60 – 64 years old (figure 4.7).

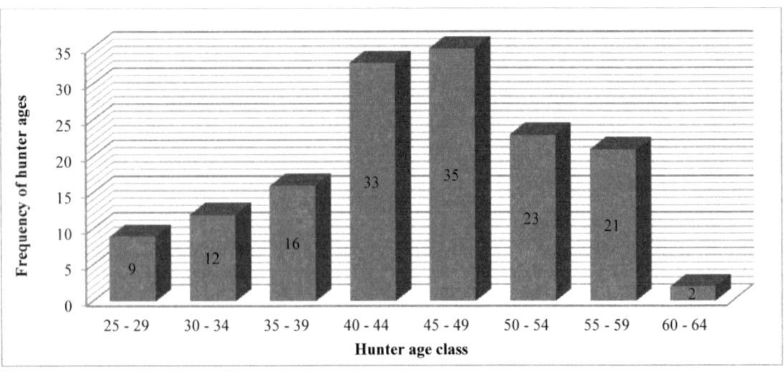

Figure 4.7 Age class of hunters from all study sites.

4.3.2. Traditional rules towards hunting *Goura victoria*

Hunters in all study sites reported several traditional regulations concerning hunting on *Goura*. These kind of traditional rules on local hunting system include taboos, restrictions and habits which usually have to be considered by each hunter (table 4.5). Some of the traditional rules were also implemented in hunting of *G.victoria*, such as the restriction for women to involve in hunting and to know about the hunting occasions. Due to certain rules, direct observation by researcher on hunting activities in the forest was not available to be done.

Table 4.5 List of taboos and ban rules on hunting found in study are.a

No.	Code	Types of rule and taboo
1.	a	It is uncomfortable for some hunters when the woman involves in their hunting team.
2.	b	When his wife is pregnant, the hunter or a hunting-team member is banned to hunt.
3.	c	There is a rule that when some hunters want to go out for hunting *G.victoria*, the plan should not be mentioned or talked by the hunters or their families.
4.	d	The hunters must ask Ondoafi's permission, as the forest land owner – sometimes including all village chiefs for administration purposes, before doing hunting
5.	e	The area of hunting is considered as the sacred area of the villagers ancestral lands, which should not be used as hunting ground
6.	f	It is prohibited to hunt due to Indonesian Government Law.

Moreover, the traditional rules and restrictions about hunting practices has become the main consideration for each hunter to pursue the hunting trip (table 4.6).

Table 4.6 Responses from Papuan hunters in each study site for the rules and taboos on hunting. (a. It is uncomfortable for some hunters when the woman involves in their hunting team, b. When his wife is pregnant, the hunter or a hunting-team member is banned to hunt. c. There is a rule that when some hunters want to go out for hunting *G.victoria*, the plan should not be mentioned or talked out by the hunters or their families, d. The hunters must ask Ondoafi's permission, sometimes including the village chiefs for administration purposes, before doing hunting, e. The area of hunting is considered as a sacred area or villagers ancestral lands, where hunting is not allowed, f. It is prohibited to hunt *Goura* due to the Indonesian Government Law).

No.	District	Study sites	Number of hunter responded to the rules on hunting *G.victoria*					
			a	b	c	d	e	f
1	Mamberamo Tengah	Buare	7	4	2	8	8	2
2	Unurumguay	Unurumguay	22	15	12	33	33	11
3	Bonggo	Bonggo	55	40	27	78	78	28
4	Supiori Utara	Supiori	15	14	15	32	32	9
Total (respondents)			99	73	56	151	151	50
Percentage (%) of each rule or taboo			65,56	48,34	37,09	100	100	33,11

The authority of tribe chief (Ondoafi) and village leader and the permissions for hunting wildlife in ancestral or customary lands have high values (table 4.6). Those authorities are still highly appreciated in local hunting system. It seems that hunters will avoid to hunting when any of those items are unfulfilled. In general, the trend of hunter responses on the rules and taboos about hunting system was similar for each study site. The rules of "asking Ondoafi's permission" and "specific hunting area" gained the highest hunter responses (table 4.6).

4.3.3. Protection of *Goura victoria*

Regarding the presence of other hunters from outside the village, all respondents agreed to the same option, that those hunters should have a permit from Ondoafi and the head of village, before they go for hunting. They also agreed with the ideas on protection of *G.victoria* from poaching by non-locals. In all study sites, around 56% of all hunters agreed to protect *G.victoria*, while the others (44%) considered it would not need any protection. For each study site, the response of hunters on protection of *Goura* is varied (figure 4.8).

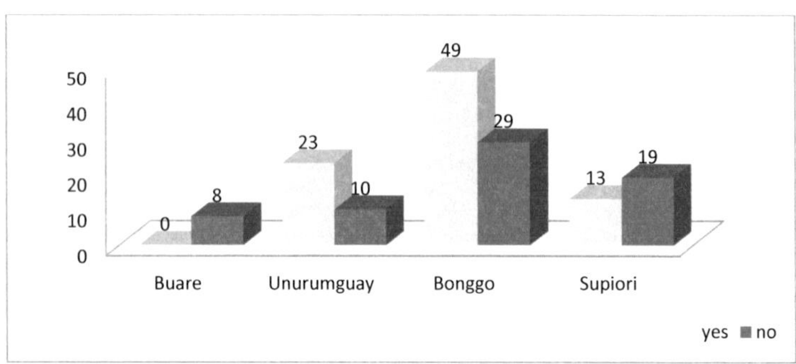

Figure 4.8 The attitude of protecting *G.victoria* in each study site.

4.3.4. Hunting attributes

Every hunter in Papua usually uses their own hunting tools such as cleaver, spear, bow and arrows, and snare. Different size of snares are usually set up to trap different animals, for instance to capture *Goura*, Megapode birds, small lizards and spiny bandicoots, they used the small size snares. Sometimes the hunters use plants that already shaped like a subtle rope, which are strong enough for making foot snares. The hunters commonly bring also other tools like knife and flashlight. About 8% of hunters in all study sites used air guns, 12% used dogs and 80% preffered to use trap snares in their hunting practices rather than using rifle gun and dogs (figure 4.9).

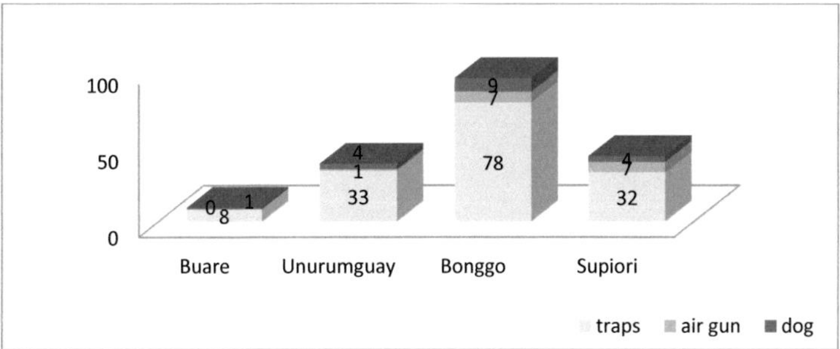

Figure 4.9 Hunting methods used by Papuan hunters.

Obviously, the figure showed that the use of air gun in Bonggo and Supiori is less frequently than that in Buare and Unurumguay. However hunting practices using dogs and snare traps were common in all study sites, particularly when the target animals included *G.victoria*. Hunting on *G.victoria* using bow and arrow rarely occurred, because the hunters should produce particular type of bow and arrow for catching this species. In fact, it is easier to capture *G.victoria* using foot snares than applying other hunting equipments.

The level of hunting frequency was higher in Bonggo area then in other study sites (figure 4.10) and the lowest was found in Buare area. Hunting practices in all sites were categorized as often activity (54%), while the type of frequently hunting and rare hunting are about 29% and 17%, respectively.

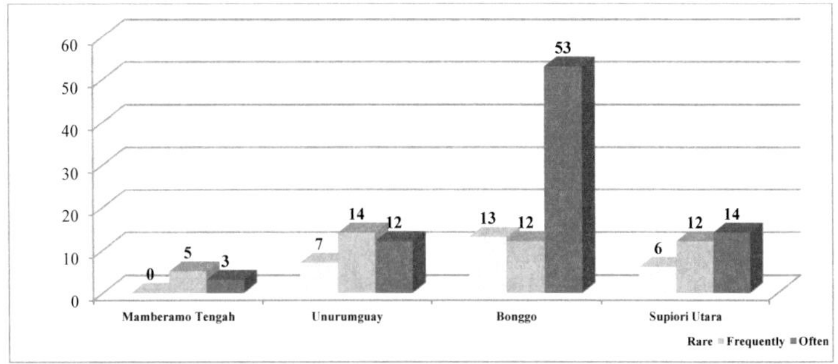

Figure 4.10 Frequency of hunting *G. victoria* by the Papuan hunter in each study site.

In regard with the season, weather and timing, the hunters generally have no specific limitation in their activities (figure 4.11). All respondents choose to hunt without dependence on particular season, although the activities were less in rainy days. Likewise, they have no specific time for hunting, but mostly (84.77%) prefer to hunt during the night until dawn (around 8 pm to 5 am) and the rest prefer to hunt in daytime or every time (15.23%).

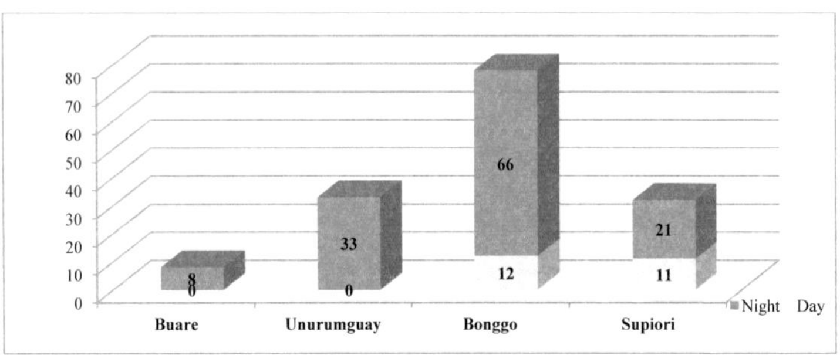

Figure 4.11 Time preferred of Papuan hunter to hunt *G. victoria* in each study site.

The hunters in all study sites tend to choose further area than the forest nearby when they go for hunting. The chance to capture animals is getting higher along with the increasing of distance from the village. Indeed, the long distance of hunting area has no effect on hunting activities, because usually hunters set up several simple rest points and they also plant some crops on their way to hunt.

Figure 4.12, shows that hunters in Buare area hunt in the middle (3-5 km) and far (>5 km) distance from their village, while hunters in three other areas choosed to hunt in three level of distance.

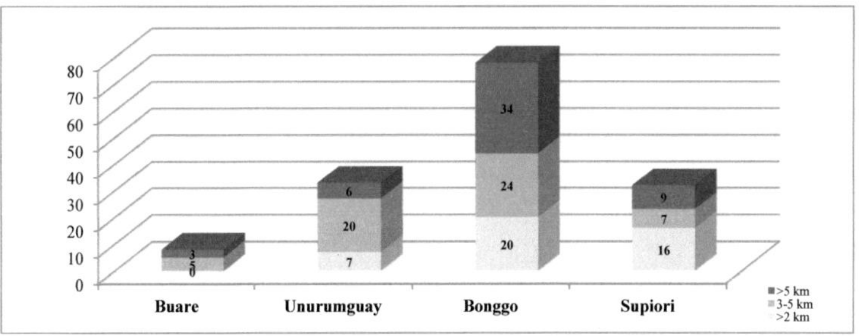

Figure 4.12 Distance to hunting grounds in each study site.

In all study sites, most of the hunters choose to hunt alone (79.47%) rather than in groups (20.53%), because hunting in group means that each harvested animal should be shared, whereas individual may gain the whole captured animal. Group and single hunting type was different among the study sites (figure 4.13).

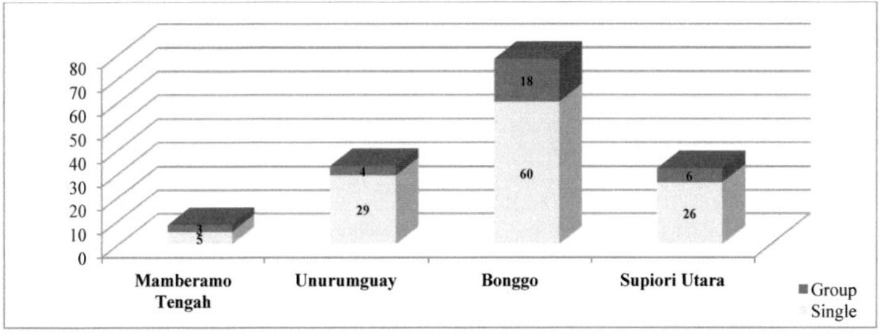

Figure 4.13 Hunter group size in each study site

4.3.5. *Goura victoria* and other hunted animals

There was a same answer from all respondents about wild animals as hunting target. All hunters basically prefer to hunt wild pig as the main target, though they would also catch *G.victoria* if they can. From all study sites, a total of 15 species of wild animals were listed as the common hunted animals. Twelve species of those animals are formally protected by Government Regulation (Law

Act No. 7/1999) and categorized as Vulnerable Species by the IUCN Red List (IUCN 2009). Several species are listed in Appendix I and II on CITES (2012) (table 4.7). All the animals listed were killed or caught for meat, but it was observed that some hunters also located and collected the eggs of megapodes and cassowary bird for sale or for consumption.

Table 4.7 List of wild animal captured and/or consumed by hunters from all study sites.

No.	Common Name	Scientific Name	Percentage of Respondent	Protection status		
				IUCN-Red List	CITES	Indonesian Law
1	Victoria Crowned Pigeon	*Goura victoria*	**64.20**	VU	II	Yes
2	Wild Boar	*Sus crofa*	**100.00**	LC	UC	No
3	Spiny Bandicoot	*Echymiphera kalubu*	**70.20**	LC	UC	No
4	Fruit-Pigeon	*Ducula* spp	57.62	LC	UC	No
5	Brown collared-brush turkey	*Talegala jobiensis*	48.34	VU	UC	Yes
6	Northern Common Cuscus	*Phalanger orientalis*	46.36	VU	II	Yes
7	Common forest Wallaby	*Dorcopsis muelleri*	38.41	VU	UC	Yes
8	Northern Cassowary	*Casuarius unappendiculatus*	37.75	VU	UC	Yes
9	Trees Cangaroo	*Dendrolagus dorianus*	34.44	VU	UC	Yes
10	Spotted Tree Monitor lizard	*Varannus similis*	26.49	VU	UC	Yes
11	Paradise birds	*Paradisaea minor*	25.17	LC	II	Yes
12	Emerald Monitor lizard	*Varanus prasinus prasinus*	20.53	VU	II	Yes
13	Deer	*Cervus timorensis*	19.87	VU	UC	Yes
14	Common Scrubfowl Cinnamon	*Megapodius freycinet*	18.54	VU	UC	Yes
15	Estuarine Crocodile	*Crocodylus porosus*	5.96	VU	I	Yes

Notes: VU: Vulnerable, LC: Least Concern, UC: unclear, I: CITES appendix I, II (Categorized based on IUCN Red List status, CITES and Law Act No.7/1999. Identification based on field guides: Beehler *et al* (1986), Coates and Peckover (2000), and Petocz (1994).

The hunters have no standard on *G.victoria*'s age while hunting. If they catch the adult birds, these were consumed or sold directly to their neighbors in the village. When they found the juveniles, although it rarely occurred, the birds would be raised up and kept as pets. Moreover, the hunters have no option in choosing specific sex of *G.victoria* when hunting. In fact, there were no obvious differences between male and female birds, and it did not matter to the hunters. The male and female can only be distinguished based on the body size. The hunters also reported that hunting on *G.victoria* could be categorized as easy hunting activity (figure 4.14).

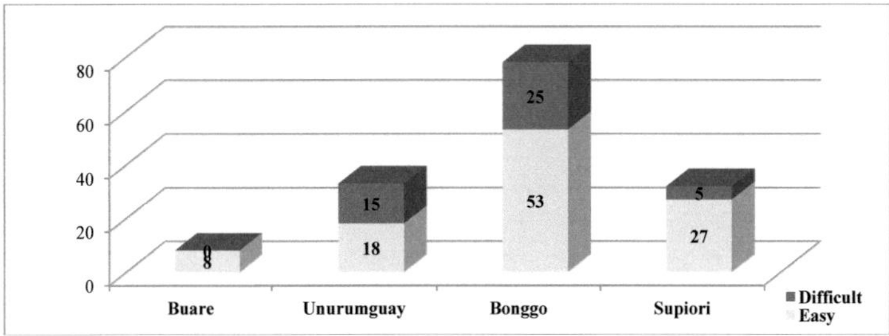

Figure 4.14 Level of difficulty on hunting *G.victoria* in each study site.

Nonetheless, the low effort in hunting does not always lead to the high catch of the bird in one trip. About 48.30% of the hunters admitted that they can catch one bird in each hunting trip and only 5.96% can catch two birds in the same trip (figure 4.15). However around 54% of the respondents in all study sites were success to catch the birds on one hunting trip while the rest (46%) were not successful. It was also revealed that the hunters in all study sites never catch more than two birds at one hunting trip (figure 4.16).

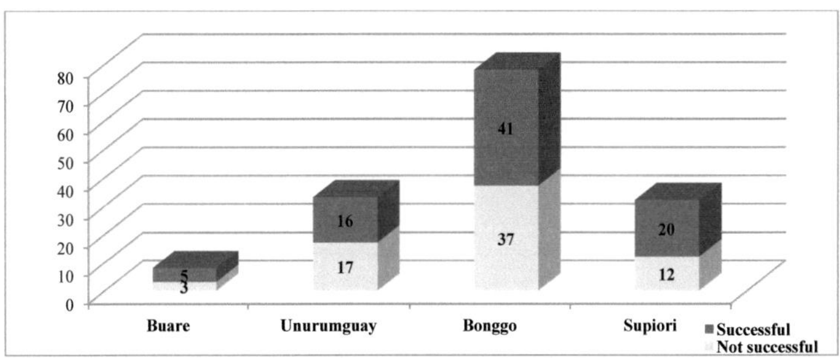

Figure 4.15 Achievement of hunting on *G.victoria* per one hunting trip in each
study site.

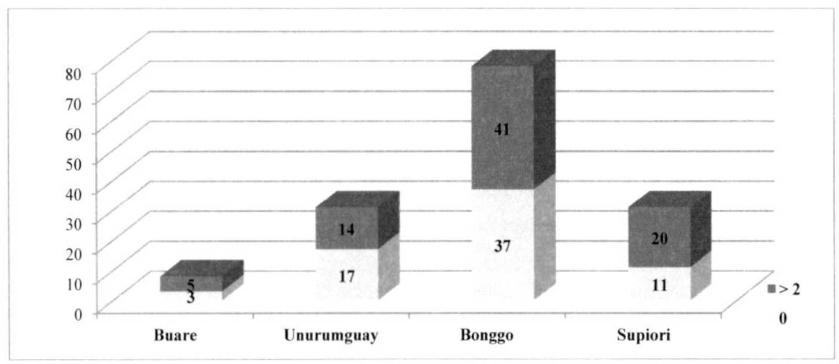

Figure 4.16 Number of *G.victoria* caught by the hunters in each study sites.

4.3.6. The use of meat of *Goura victoria*

It is locally known that the main reason of hunting on *G.victoria* is to catch the live birds for gain fresh money to fulfill the daily necessities of hunter and his family. The *Goura* meat was consumed, if the birds found injured or dead on hunting ground. In all study sites around 24.50% of hunters reported they consumed *Goura* meat. The consumption of *Goura* meat by the hunters are classified into three levels (rarely, occasionally and commonly), and these levels are different among the study sites (figure 4. 17).

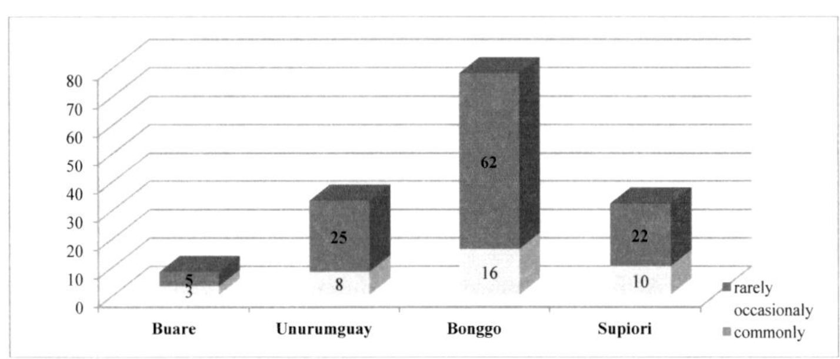

Figure 4.17 Level of consumption on *Goura* meat in each study site

The hunters prefer to catch *G.victoria* alive, because its price is higher than a piece of *Goura* meat. Overall, the price of one live bird of *G.victoria* is around US$12.5, and the average price of a piece smoked *Goura* meat is around US$ 2. Both the live birds and smoked *Goura* meat were sold around their village or to their neighbor, or bartered (table 4.8). Live birds can be exchanged with essential goods such as the packets of salt, sugar, coffee, and even cigarettes, or can be purchased as pets. There was also another case in the Supiori site, where the live *Goura* was exchanged with several basic needs, and also bullets or fishing equipments like nylons rope and a pack of hook, metal ballast, or with certain fishing nets.

Table 4.8 Selling price of *G.victoria* in each study site

No.	District	Price (in US $)		
		Live bird	A piece of meat bird (± 1 kg)	In barter system
1.	Mamberamo Tengah	5	1 – 2	5
2.	Unurumguay	10 –15	1.5 – 3	5 - 10
3.	Bonggo	10 –20	1.5 – 3	5 –10
4.	Supiori Utara	10 –15	-	5 –15

Note: 1US$ equal to ID Rp. 10.000,-

4.3.7 Relationship between numbers of *Goura victoria* caught by the hunters in each study site with the variables of hunting practice

Two non-parametric statistical analyses (Kruskal-Wallis test and Mann-Whitney U test) were performed to determine the relationship between numbers of *Goura* captured and several hunting variables (followed the structure from Aiyadurai *et al*, 2011). Hunting variables in these analyses include hunting distance, hunting group size, hunting frequency, frequency of *Goura* meat consumption, and numbers of hunting using air gun and using dogs. Both analyses were carried out and performed separately for each study site.

The results of Kruskal-Wallis test showed that there was no relationship between numbers of *Goura* captured with six variables of hunting practices in Buare (table 4.9). This test obtained the value of $P_{test} > P_{0.05}$, and produced the similar result as from Mann-Whitney U-test.

Table 4.9 The relationship between numbers of *G.victoria* captured and six hunting variables in Buare site.

Variables	Statistics test						
	Kruskal-Wallis test (n= 8)				Mann-Whitney U test (n= 8)		
	χ^2	df	Ptest	Result	Z	Ptest	Result
Hunting distance	2,52	1	0,112	ns	-1,587	0,112	ns
Hunting group size	2,52	1	0,112	ns	-1,587	0,112	ns
Hunting frequency	2,52	1	0,112	ns	-1,587	0,112	ns
Frequency of *Goura* meat consumption	0,031	1	0,86	ns			
Frequency of hunting using air-gun							
Frequency of hunting using dogs	0.60	1	0.439	ns	-0,775	0,439	ns

Notes: ns: no significant (χ^2 test>$\chi^2\alpha 0.05$); (Ztest>Z$\alpha 0.05$); black box: the variable cannot be analyzed with the statistics test-SPSS 17

The results from Supiori were slightly different than that from Buare. Two hunting variables showed relationship with the numbers of *Goura* captured in Supiori (table 4.10). Statistically, the frequency of consumption on *Goura* meat and frequency of hunting using air gun showed significant relationships with the number of *Goura* captured (χ^2=0.012 P<0.05, χ^2=0.000P<0.05), respectively (table 4.10). However, the Z test only showed the significant relationship between frequencies of hunting using air gun with the number of *Goura* captured (Z=0.000 P<0.05).

Table 4.10 The relationship between numbers of *G.victoria* captured and six hunting variables in Supiori site.

Variables	Statistic test						
	Kruskal-Wallis test (n= 32)				Mann-Whitney U test (n= 32)		
	χ^2	df	Ptest	Result	Z	Ptest	Result
Hunting distance	0,102	1	0,749	ns			
Hunting group size	0,102	1	0,749	ns	-0,320	0,749	ns
Hunting frequency	1,433	2	0,486	ns	-0,522	0,602	ns
Frequency of *Goura* meat consumption	**6,313**	**1**	**0,012**	*			
Frequency of hunting using air-gun	**12,400**	**1**	**0,000**	*	**-3,521**	**0,000**	*
Frequency of hunting using dogs	0,360	1	0,568	ns	-0,571	0,568	ns

Notes: ns: no significant (χ^2test>$\chi^2\alpha 0.05$),(Ztest>Z$\alpha 0.05$), and*:significant (Ptest<P$\alpha 0.05$); black box: the variable cannot be analyzed with the statistics test-SPSS 17

The results from Unurumguay are similar to those from Buare. The statistical test showed no relationship between all variables tested with the number of *Goura* captured (table 4.11).

Table 4.11 The relationship between the number of *G.victoria* captured and six hunting variables in Unurumguay site.

Variables	Statistics test						
	Kruskal-Wallis test (n=33)				Mann-Whitney U test (n= 33)		
	χ^2	df	Ptest	Result	Z	Ptest	Result
Hunting distance	0,299	2	0,861	ns	-0,434	0,665	ns
Hunting group size	0,004	1	0,949	ns	-0,064	0,949	ns
Hunting frequency	1,283	2	0,526	ns	-0,203	0,839	ns
Frequency of *Goura* meat consumption	0,805	1	0,370	ns			
Frequency of hunting using air-gun	1,063	1	0,303	ns	-1,031	0,303	ns
Frequency of hunting using dogs	1,242	1	0,265	ns	-1,115	0,265	ns

Notes: ns: no significant (χ^2 test>$\chi^2\alpha0.05$), (Ztest>Zα0.05) and black box: the variable cannot be analyzed with the statistics test-SPSS 17

Results from Bonggo were found different than those in other study sites (table 4.12). In Bonggo, the size of hunting group and hunting practice using air gun have significant relationship with the number of *Goura* captured.

Table 4.12 The relationship between the number of *G.victoria* captured and six hunting variables in Bonggo site.

Variables	Statistic test						
	Kruskal-Wallis test (n= 78)				Mann-Whitney U test (n=78)		
	χ^2	df	Ptest	Result	Z	Ptest	Result
Hunting distance	3,494	2	0,174	ns	-0,045	0,964	ns
Hunting group size	**5,517**	**1**	**0,019**	*	**-2,349**	**0,019**	*
Hunting frequency	3,934	2	0,14	ns	-1,517	0,129	ns
Frequency of *Goura* meat consumption	0,790	1	0,374	ns			
Frequency of hunting using air-gun	**18,119**	**1**	**0,000**	*	**-4,257**	**0,000**	*
Frequency of hunting using dogs	2,312	1	0,128	ns	-1,521	0,128	ns

Notes: ns: no significant (χ^2 test>$\chi^2\alpha0.05$), (Z $_{test}$>Z$\alpha_{0.05}$) and *: significant (P $_{test}$<P$\alpha_{0.05}$); black box: the variable cannot be analyzed with the statistics test-SPSS 17

The results from all study sites showed that hunting practice using air gun, the frequency of meat consumption and the size of hunting group may affect the population of *G.victoria*. The other variables including the distance of hunting area, the level of hunting frequency and the practice of hunting using dogs, have apparently no significant effect to the bird's population in all study sites (table 4.9, table 4.10, table 4.11and table 4.12).

A multiple linier regression analysis was carried out to figure out which hunting variables can mostly influence the number of *Goura* captured in each study site (table 4.13, table 4.14). The results of Analysis of Variance (ANOVA) and regression analysis showed that some hunting factors have significant influences on the number of *Goura* captured in Supiori and Bonggo.

Table 4.13   The result of analysis of variance test on regression analysis between five hunting variables (**) and the number of *G.victoria* captured in each study site.

Study site	Result of ANOVA on multiple regression analysis					
	F test	Result	Conclusion	P test	Result	Conclusion
Buare	1.406	$F_{test}>F_{(2,5)\alpha0.05}$	ns	0.328	$P_{test}>P_{\alpha}0.05$	ns
Supiori	6.775	$F_{test}<F_{(5,26)\alpha0.05}$	*	0.000	$P_{test}<P_{\alpha}0.05$	*
Unurumguay	0.374	$F_{test}>F_{(4,28)\alpha0.05}$	ns	0.825	$P_{test}>P_{\alpha}0.05$	ns
Bonggo	10.262	$F_{test}>F_{(4,73)\alpha0.05}$	*	0.000	$P_{test}<P_{\alpha}0.05$	*

Notes: ns: no significant; *: significant, ** five hunting variables are: Distance when hunting (x1), hunting group size (x2), using air-gun (x3), using dogs (x4) and frequency of *G.victoria* meat consumption (x5).

Furthermore, table 4.14 showed that most hunting variables have no significant effect on the number of *Goura* captured in each study site. However, the variable of hunting using air gun can be considered as the most factor influencing the number of *Goura* captured in Supiori and Bonggo.

Table 4.14 The result of multiple linear regression between the numbers of *G.victoria* captured and five hunting variables in each study site.

Variables of Multi Linier Regression	Study site			
	Buare	Supiori	Unurumguay	Bonggo
Constants (C)	-0,2	3,46	1,56	1,76
Distance when hunting (X1)		-0,01	-0,02	0,15
Hunting group size (X2)	0,6	0,21	0,08	0,05
Using air-gun (X3)		-1,28*	-0,37	-1,04*
Using dogs (X4)	-3,2E-17	-0,24	-0,21	0,38
Frequency of *G.victoria* meat consumption (X5)		-0,36		
R^2	0.36	0.57	0.05	0.36

Notes: *: significant ($P_{test}<P\alpha_{0.05}$), black box: the variable cannot be analyzed with the statistics test-SPSS 17

The results from regression analysis showed that hunters in Buare caught the least number of *G.victoria*, but this is not statistically significant compared with the hunters in other study sites ($R^2 = 0.36$, see in table 4.14).   This value was confirmed that the correlation between variables can be classified as modest correlation.   Supported by the value of variables hunting group size and hunting using dogs, these variables are only influential altogether for the level of 36% on the number of *Goura* captured, while the 64% was likely affected by other factors.

The results from Buare were slightly different from those found in Supiori, where all tested variables were almost close to the zero value.   Statistically, only variable of hunting using air gun showed significant effect on the number of

Goura captured. The hunters in Supiori site caught the most birds among all study sites (approximately 3.46 individuals). This significant result means that the increase of hunting practices using air gun may increase the number of captured bird (F = 6.775, P<0.05, see table 4.13).

The results from Unurumguay were different than those from other study sites. It showed that the number of *Goura* captured was affected simultaneously by all variables except the frequency of consumption on *G.victoria's* meat (F= 0.825, P>0.05). The multiple linier regression analysis (table 4.14) showed the weak correlation between Buare, Supiori and Bonggo area (R²= 0.36, 0.57 and 0.36), respectively. It means that hunting success on *G.victoria* also influenced by other factors than hunting variables tested. Compared with other study sites, the hunters in Supiori and Bonggo can be classified as active hunters (F= 6.775, P<0.05 and F= 10.262 P<0.05), see also table 4.13, because the number of *Goura* captured increased significantly by the raise of hunting using air gun.

4.3.8. Estimation of hunting sustainability and impact of harvesting on *Goura victoria*

Calculation and estimation of the maximum sustainable harvest of *G.victoria* were done using the demographic information from captivity and literatures (table 4.15).

Table 4.15 Demographic information from captivity and literature of *G.victoria.*

Variabels	The values	References and Notes
Age of first reproduction (**a**)	5 years	Beltermann and Poot (2008)
Age of last reproduction (**w**)	15 years	Beltermann and Poot (2008)
Reach sexual maturity	1.5 years	Beltermann and Poot (2008)
Lay first egg at an age	1.5 - 2 years	Beltermann and Poot (2008)
Estimation on generation time	2 - 8 years	Beltermann and Poot (2008)
Incubation time	20-30 days	Beltermann and Poot (2008)
Age of chicks leaving the nest	13 weeks old	Gibbs *et al* (2001), Baptista *et al* (1997)
Mean weight or body mass	±2.5 kg	Pangau-Adam and Noske (2010)
Life expectancy	**0.2** (long-live species)	Robinson and Redford (1994), Bennett and Robinson (2000), Robinson (2000)
Numbers of egg per years (most in dry season or the end of wet season)	1	Coates (1985), Gibbs *et al* (2001)
The value of annual birth rate of female offspring (**b**)	**0.097**	Based on Noss (2000)

Based on data in table 4.15, the value of the maximum intrinsic rate of increase of a population that is not limited by food, space, resource competition, or predation (**r$_{max}$**) is 0,001, and the value of the maximum finite rate of increase

($\lambda_{\mathbf{max}}$) is 1,001 for *Goura victoria*. By using the value of λ_{max}, the value of $P_{max(D)}$ or the maximum annual production of *Goura* at observed density in each study site, can be calculated. Furthermore, it can also be calculated the values of maximum sustainable annual harvest, maximum sustainable harvest based on hunting area size, and the estimation of current annual harvest within the hunting area (table 4.16).

Table 4.16 Estimation of maximum annual production at this density within hunting area size or $\mathbf{P_{max(D)}}$, estimation of maximum sustainable annual harvest levels within hunting area size and estimation of current annual harvest within the hunting area size for *G.victoria* in each study site

Study site	Population density (D) of *G.victoria* (**ind/km²**)	Hunting area size (**km²**)	Population estimation within hunting area size (**individual, 95% CI***)	Estimation of maximum annual production at this density within hunting area size or Pmax D (**individual**)	Estimation of maximum sustainable anual harvest within hunting area size (**individual**)	Estimation of current annual harvest within the hunting area size** (**individual**)
Buare	41.74	78.5	2,451 - 4,380	1.97	0.39	83.20
Supiori	40.29	113.04	3,344 - 6,206	2.73	0.55	332.80
Unurumguay	30.8	113.04	2,589 - 4,682	2.09	0.42	343.20
Bonggo	13.1	314	3,266 - 5,178	2.47	0.49	811.20

Notes: * 95% Confidential Interval; ** assumed that hunters only hunted *G.victoria* in the given hunting area

The value of estimated current harvest in each hunting area was ranging between 83.20 to 811.20 individual of *Goura victoria* while the value of estimated maximum sustainable harvest per each hunting area are between 0.39 to 0.55 individual (table 4.16). These values show that the estimated current harvest levels of *Goura victoria* in each hunting area were obviously higher than the estimated maximum sustainable harvest in the same hunting area. This may indicate that the hunting activities by Papuan hunters is not sustainable and have negative effects on the population of *Goura victoria*.

CHAPTER 5: DISCUSSION

5.1 Forest area as the habitat of *Goura victoria*

Both Papua and West Papua Provinces have approximately 42 million hectares of forests that cover about 80% of the lands. From that area, 28 million hectares lies in Papua province and is classified into several forest types; production forest, limited production forest, permanent production forest and protection forest (BPS Papua 2010[b]).

Currently the Papuan forests are threatened through forest exploitation such as logging activities, establishment of new districts and a variety of forest conversion for agricultural purposes like oil palm plantation and land opening for transmigration land. In addition, slash and burn farmland practised by local peole has allegedly involved in changing the forest composition and structure in lowland forest of Papua. This also occurred in all sites of current study.

Based on the results, floristic composition in Buare forest was dominated by *Pimeliondendron amboinensis Hassk* (IVI=38.53%), *Canarium indicum* (IVI=34.58%) in Supiori forest, *Pimeliondendron amboinensis Hassk* (IVI=40.59%) in Unurumguay forest and *Syzygium* sp (IVI=35.25%) in Bonggo forest. In this study area, the species from family of Euphorbiaceae were found in all of study site, like in other places in Papua (Mirmanto 2009, Kabelen and Warpur 2009), West Java (Purwaningsih and Yusuf 2008), and Mentawai-Siberut (Hadi *et al* 2009). Euphorbiaceae is considered as the most common tree family in secondary forest. This plant family has highly adaptability in different environment conditions specifically in the lowland forest (Purwaningsih and Yusuf 2008).

The forest structure in each study site was performed by vegetation from mid-lower canopy class (strata B: Richards 1996), dominated by trees under 25 m-height and mostly had *dbh* less than 35 cm. This situation was similar to the vegetation structure in Tangkoko Nature Reserve in North Sulawesi Indonesia, which was merely dominated by the trees with diameter breast height-dbh ranging from 21.1-26.6 cm (Rosenbaum *et al* 1998). It was also comparable with vegetation in primary forest in Siberut-Mentawai, Sumatera, Indonesia, which dominated by the trees with diameter less than 40 cm and the total height under 20 cm (Hadi *et al* 2009). The vegetation structure in all study sites and in other forest described above can be categorized as unusual pattern of vegetation structure in lowland tropical forest.

The lowland forest located near to villages has become more accessible to humans, and this may lead to the high exploitation of the large trees. As the result, the vegetation remnants are dominated by the lower canopy trees (<20 m). This pattern was also occurred in Siberut-Sumatera, where the big trees in lowland forest were fell down for boats and house construction (Hadi *et al* 2009). Vegetation in all study sites were dominated by the trees with small diameter and lower canopy. In Unurumguay site, forest vegetation around the village was used as shade-stands for cocoa crops (*Theobroma cacao*), whereas in Buare, Supiori and Bonggo sites, forest areas adjacent to the settlement were used as farmland planted with several crops like taro, cassava, banana and vegetables. Additionally illegal logging activity for house construction and gathering fuel wood were practiced in all sites. This forest use contributed to the changes of vegetation structures in the study area.

Forest destruction related to changes of habitat quality, forest structure and composition usually leads to affect the animals including birds that inhabit the forest. For instance, logging activity can cause forest damages including the loss of food trees and nesting trees, and provide more access to the remote and undisturbed forest area. This has also happened in several forest areas in Indonesia and other parts around the world. For instance, habitat destruction due to logging activity has become a major threat to the existence of primates in Bacan Island (Rosenbaum *et al* 1998) and on Cracids population in Peru (Barrio 2011). Unsustainable logging practice and small scale of illegal logging in Seram Island might destroy the forest canopy and reduce habitat of Seram Cockatoo, specifically species of nesting trees and food resources trees (Kinnaird *et al* 2003). Some anthropogenic pressure such as uncontrolled harvesting of non-timber forest products, forest logging and forest fire have threatened the population of the Red-knobbed hornbills in Buton Island (Winarni and Jones 2012). The decrease of fruit tree quantity can be followed by the shrink of forest quality for frugivores

(Marthin and Blackburn 2012). One of the evidence is the case with the Red-knobbed hornbill, a frugivore in Buton Island, which is so sensitive to the forest disturbance (Winarni and Jones 2012). This bird will move much further into the forest to find their food trees, because their previous habitat already changed. Information on forest structure in this study showed that forest structure with domination of trees from lower canopy allegedly influenced the presence of *G.victoria*. The bird is specialized forest floor dweller, but still need the good forest like a strong branch of tree for perching, laying nest or hiding from predators. It is observed in this study, that *Goura* was flying away from disturbed forest to find the other forest area with a better quality.

There are some requirements for a forest area to be considered as a suitable habitat for *Goura* spp. These factors are including sufficient light, availability of small rivers, wet forest floors, mud pools, shallow marshes, and also food resources (Setio and Lekitoo 2000). Particularly, the presence of food trees and dominant species of food trees may affect the number of wildlife species within a certain forest area. Recent study documented that there were at least nine tree species considered as the common food trees for Papuan birds: *Myristica* sp, *Eugenia anomala*, *Pometia* sp, *Cananga odorata*, *Canarium asperum*, *Pimeliodendron amboinicum* Hassk, *Intsia* sp, *Ficus* sp and *Chisocheton* sp (Setio and Lekitoo 2000). Likewise, the presence of species like *Syzigium* sp, *Arthocarpus* sp, *Terminalia* sp and *Ficus* sp may provide sustainable food source for birds and this may become a good indication for the potential of conserving birds and other animals (Alamgir *et al* 2011).

This *Goura* study showed that the forest area in Buare, Supiori, Unurumguay and Bonggo were rich in food trees for birds. Several food tree species such as *Pimeliodendron amboinicum* Hassk, *Intsia* sp, *Pometia* sp, *Canarium* sp, *Eugenia anomala* and *Myristica* sp dominated in all study sites (table 4.3, appendix 1, 2, 3, 4 and 5). Moreover, another study also found that tree species of *Ficus* sp, *Cananga odorata*, *Canarium australicum*, *Litsea* sp, *Eugenia* sp, *Syzigium* sp, *Planchonella firm* and *Vitex cofasus* were considered as favourite food trees for fruit-pigeons specifically those from genus *Ducula* (9 species) and *Ptilinopus* (4 species) in New Guinea (Frith *et al* 1976). The later study reported that the birds of New Guinea prefer to eat fruits and berries. However more studies are still needed to assess food preference by frugivorours birds in New Guinea. Analysis on forest composition in all study sites found that about a total of 51 species are being the fruit trees. This is about 36% from all species encountered and belong to 14 families (14% from all families encountered). This result indicated that forest area within each study site has potential food sources for frugivorous and generalized birds, although there was unclear information on the favorite fruits for *Goura*. In all study sites, all information about food sources of *Goura* originated from traditional knowledge of the local hunter. They reported that the bird feeds on fruits of *Palaquium amboinensis*, *Canarium* spp, *Terminalia* spp, *Ficus* spp, and also feeds on the stem sap of iron wood (*Intsia* spp). All these trees were encountered in all study sites (appendix 1, 2, 3 and 4). However, further detailed research about the fruit tree species that were consumed by *G.victoria* is still needed.

Based on the local knowledge of hunters on fruit trees species as food source of *Goura victoria*, combined with the information from Snow (1981) and Frith *et al* (1976), the number of food trees can be counted. There were 19 tree species of 12 families in Buare forest, 21 species of 12 families in Supiori forest, 19 species in 10 families in Unurumguay, and 12 species in 11 families in Bonggo forest (appendix 5). Furthermore, another study on feeding trees stated that there were around 142 species of trees listed as food sources for generalized and specialized frugivorous birds including family Columbidae in Australasia Region. This list also contained some tree species that found in the study area, such as *Myristica* sp, *Planchonella* sp, *Eugenia* sp, *Syzygium* sp, *Terminalia* sp, *Canarium* sp, and *Vitex* sp. Fruits from these species are prefered by eight frugivorous bird families in New Guinea region (Snow 1981). These fruit trees were distributed abundantly in different locations in Papua. For instance, the swamp areas in all altitudes in Papua were dominated by *Syzigium* sp, *Garcinia* sp, *Canarium* sp, *Myristica* sp, *Terminalia* sp and *Eugenia* sp, while the lowland forests were dominated by *Terminalia* sp, *Myristica* sp, *Garcinia* sp, *Syzigium* sp, *Planchonella* sp, and *Canarium* sp (Johns *et al* 2007[a], 2007[b]).

Although in this study area there was only around 35% of feeding tree species compared to Australasia fruit plant species list from Snow (1981) and about 67% of total tree families from Frith *et al* (1976), the list of trees as food source from this study can become the preliminary information about food source of Papuan frugivorous birds including *G.victoria*. The presence of *G.victoria* in each study site is presumably influenced by the abundance and dominance of fruit trees.

Goura victoria is very sensitive to extinction risk, specifically in the early stage of habitat loss, because this species required forest trees for foraging. This response of *Goura* can be used as indicator of forest degradation (King and Nijboer 1994, Castelletta *et al* 2000, Bird Life International 2012). The similar character also applied to the large-bodied game birds that become particularly sensitive to habitat loss and forest fragmentation (Thornton *et al* 2012). It was reported that the *Buceros rhinoceros*, a large-bodied forest dweller bird, was avoiding disturbed forest because of the loss of food tree sources (Anggraini *et al* 2000).

Overall, there are many important factors interrelated and driving a bird species in responding forest destruction like logging and forest fragmentation, (Cleary *et al* 2007). These factors consist of forest composition and structure, mainly vertical structure (Cleary *et al* 2007), bird's body size (Cleary *et al* 2007, Marthin and Blackburn 2012), and bird specialization, for instance insectivores or frugivores, which are more prone to the extinction (Cleary *et al* 2007).

All species of crown pigeon are well-known as frugivores (Beehler 1982, Beehler *et al* 1986), though they can also feed on worms, small insects and even small shellfish from the beach or muddy-river banks opportunistically (Gibbs *et al* 2001, Baptista *et al* 1997). In general, the information on specific food tree species for *G.victoria* is still very poor. Therefore, the inclusive information about food sources for *G.victoria* in their habitats is still required and has to be documented, specifically due to the peril of habitat loss to the bird's population (Bird Life International 2012).

5.2 The population of *Goura victoria* in study sites

This study showed that the population densities of *G.victoria* are slightly different between each study site. The estimation was done withhe assumption that all lowland forest in all study sites can provide equal adequate habitat for *G.victoria*. Density estimation of *Goura* population in Buare-Mamberamo was higher than in Bonggo, but almost similar to the density population of *Goura* in Supiori and Unurumguay. The density estimation in Buare was about 41.8 birds.km^{-2} in 78.5 km^2 hunting area size, it means there were approximately 2451 – 4380 birds that inhabit the hunting area of Buare forest. This prediction was merely done for Buare area only and it could not be applied to the entire forest area in Mamberamo Raya Regency. This forest is located within Mamberamo vast wathershed and has only low pressures or threats from local people such as traditional hunting on *G.victoria*. Therefore the density estimates was considerable high. Additionaly, the human population size is low and the main target animals of wildlife hunting were wildboars (Chamberlain *et al* 2004, Richard and Suryadi 2002, Mack and Alonso 2000). However, as happened in other part of Papua, rapid development including the establishment of new regencies, districts and villages are threatening the Buare forest. As the results, deforestation is increasing, and this may adversely affect the presence of Papuan wildlife (Frazier 2007). The other threat on Buare forest and *Goura* population is the regional plans to build up a large dam in Mamberamo watershed and to construct the highway of Trans Papua (Anggraeni 2007). Dam construction is considered by the government of Papua Province as an important development for local people welfare. However, it is considered, that the construction may have negative impacts on biodiversity in the forest area within Mamberamo watershed, especially the impact on *Goura* population. Actually the large dam construction is still in the planning level, but few samples of large dams in United States, China and Thailand already show the negative impacts of large dams on major rivers and on the wildlife along the rivers and watershed (McAllister 2001). Some further negative impacts might include the damage or loss of food resources and other important habitat components such as nesting sites and breeding sites of wild animals. The other serious threat is the logging concession which is still active in

Mamberamo Raya Regency (Anggraeni 2007, Elsham Papua 2008). Increasing of logging activity in Mamberamo may lead to the habitat loss of *Goura victoria.*

Supiori site has quite high population density of *G.victoria* compared to that in Unurumguay and Bonggo. It was estimated that Supiori area has about 40.3 birds per km^2 with population size of 3,344 to 6,206 birds in 113.04 km^2 hunting area size. Habitat decrease may be accounted as the key pressures on *G.victoria* in Supiori forest. The forest areas are converted to the new infrastructures, settlements for new districts and villages, road constructions as well as human activities such as shifting cultivation, collecting firewoods, felling trees for household needs, and hunting. All of these activities may increase the pressures on wildlife animals in Supiori forest.

The density estimates in Unurumguay was 30.8 birds per square kilometer and the studied hunting area was about 113.04 km^2. It could be predicted that *Goura's* population in in this site was about 2,589 to 4,682 birds. In Bonggo, the density estimates was 13.10 birds per square kilometer and the studied area was abot 314 km^2 so the population of *Goura* in this site was about 3,266 to 5,178 birds.

The lower population sizes of *Goura victoria* were possibly happened due to the high intensity of forest opening in the past, even before Jayapura Regency was expanded into two new regencies. The Regency of Jayapura (where Unurumguay District lies) and Sarmi Regency (where Bonggo District also lies) are located in an already developed and opened area. Sarmi Regency was formerly a district of Jayapura Regency, and has already opened its forest area to develop the infrastructures for implementing a number of government programs. These programs included transmigrations, resettlements, road constructions, and forest concessions (Anggraeni 2007). In addition, there are recently at least four large-scale plantations and two logging concessions in Jayapura Regency, while Sarmi Regency has some plantation companies that were newly developed like cocoa and coffee plantation and another three forest concessions, and three new concessions are waiting to begin their operations (Anggraeni 2007, BPKH 2008, Pangkali 2011-*in prep*). There are also small timber's companies on the community forest that managed by the community leader (Ondoafi). Beside logging and plantations, particular human activities around and within the forest area such as hunting practices and regular shifting cultivations might threaten the presence and the population of *Goura victoria.*

Forest quality is the most important feature in relation to habitat conditions and the presence of wildlife (Alvard 2000). Destruction or degradation of the forest in various forms may actually influence the presence and population of wildlife conditions, as it happened on *Goura victoria.* Particularly, Indonesia has already turned into a country with the highest level of tropical forest vanished (Corlett 2009). The occupancy of several logging companies and plantations in Papua showed the increasing deforestation in this region.

In recent days development in the northern part of Papua occur rapidly and vastly. This rapid development is driven by the government policies and human population growth including transmigration programs and the new resettlements program of Regency government for the local people. This might result in considerable pressure to the existence of rainforests in Papua. In addition, development of logging activities and oil palm plantation expansion in the northern Papua remains ongoing and is expected to affect the population of *Goura victoria.* Kinnaird *et al* (2003) reported that unsustainable logging practice in the lowland forest of Seram, Moluccas was damaging forest canopy and reduce the habitat area of Seram cockatoo (*Caccatua moluccensis*). Logging operation may be the main cause of the loss of food trees and nesting trees in natural habitat of *Goura.* Due to the high logging pressure, the nesting trees and food trees were rarely found in Bonggo forest. This may seriously affect the population abundance of *Goura.* Several studies showed that logging activity affected abundance, diversity and density of birds and primates (Johns 1983 and 1985, Wilson *et al* 1983, Marsden 1992, 1998, Waltert *et al* 2002, Marsden and Pilgrim 2003).

Other examples are the population decline of mammals and birds in Sulawesi (Riley 2002, Rosenbaum *et al* 1998, O'Brien and Kinnaird 1996). A study in Guatemala forest showed that big birds like the large Galliforms and Tinamou (Cracidae, Phasianidae and Tinamide) are very sensitive to habitat loss and forest fragmentations (Thornton *et al* 2011).

Logging might also has a great effect on the forest sustainability. Logging activities can cause the loss of keystone animals for maintaining several functions in the tropical forest such as seed dispersers and pollination agents (Robinson *et al* 1999). CIFOR studies also noted that in other part of Mamberamo, Papua, local people had already difficulties to find wild animals like paradise birds, parrots and cassowaries, due to high intensity of forest utilizations (Boissiére *et al* 2007, Padmanaba *et al* 2012), mainly through the excessive illegal logging and local timber companies.

Setting aside new infrastructure for regional development in Papua is amongst the major threat of forest degradation in Papua. Road constructions through forests area usually affect on forest loss, species pattern alteration, and more human access and disturbance in remote area (Petocz 1978, Seiler 2001). Furthermore, these roads may disrupt horizontal natural process, change landscape patterns and reduce biodiversity. Other possible impacts of road constructions can include habitat changes and direct effects on distribution and abundance of plant and animals along the roads (Geneleti 2003). In all study sites, it is observed that road construction, using skid for logging consession and forest utilization by the locals may have negative affects on the *Goura victoria* population.

Overall, this study may show that *Goura victoria* living in the lowland rainforest of the northern Papua is probably not within the peril of extinction. The population size estimated in each study site was above the population data reported by the Bird Life International (2012), which predicted that the current population of *G.victoria* is around 2,500-9,999 birds or equals with 1500-7000 mature individuals in the lowlands of northern Papua. However, the comprehensive overview on the population of *G.victoria* is difficult to be carried out, since there are insufficient studies on this subject.

5.3 Hunting of *Goura victoria* by Papuan People

5.3.1 Hunting activities in Papua-Indonesia

There are many different reasons for doing hunting, although the main reason is to fulfill family nutrition (Bennett and Robinson 2000[a], Smith 2005). For instance, wildlife meat can provide the need of animal proteins and fats for local people in Amazon region (Bennett and Robinson 2000[a], Towsend 2000), for local villagers in Sulawesi and Kalimantan, Indonesia (Alvard 2000, Lee 2000, Wadley *et al* 1997, Wadley *et al* 2004), and for local communities in New Guinea as well (Pattiselano 2003, Johnson *et al* 2004, Cuthberth 2010). The other reason of hunting is related with economic needs.

In this study, the main reason in hunting *Goura* is to gain the fresh money for fulfilling daily needs of hunter's family. *Goura victoria* is known as the high-valued bird, which can be sold alive or in the form of smoked meat. The hunters and their family sometimes consume the bird's meat as protein source, although it might be occurred very rarely.

It is similarly with other local people or hunters from other areas, who practise hunting and sell the harvest animals either dead or alive to get fresh money (Dwyer 1974, Clayton and Milner-Gulland 2000, Pattiselano 2003, Sada 2005, Smith 2005, Mahuse 2006, Pangau-Adam and Noske 2010, Jepson *et al* 2011). However, the hunters in this study did not rely entirely from the hunting practices. They can change their activities in certain times and find other jobs with quick money or greater cash whenever hunting yield seems unprofitable. Sometimes the hunters tend to be opportunistic and change their activities to other jobs for immediate needs (Jorgenson 1995, Smith 2005). This opportunistic hunter also appears on their activities in gathering forest fruits and wild vegetables, or in harvesting rattan for housing needs, and collecting areca nut and bettles when across abandoned farms, adjacent to the hunting grounds. Similarly, local people in Panama showed this manner that they do land opening and clearing, collecting medicinal plants or just pacing from one hunting site to others when go hunting (Smith 2010).

Moreover, hunting practice also has socio-cultural reasons, when it is carried out for fulfilling more than basic needs previously mentioned (Benneth and Robinson 2000[a]). This reason included animal's function as: 1) private collection to show hunters' status or pride, for example through collections of skulls, feathers, horns, leathers, claws and preserved animals (Kwapena 1984, Shaw 1969, Petocz 1978, Aiyadurai 2011); 2) and high reputation or high position in tribe hierarchy, that can be achieved through hunters ability and success. The last function can also lead to obtain a kind of competency to marry a woman (Kwapena 1984, Jorgenson 1995, Pattiselano 2003, Bird and Bird 2008, Aiyadurai *et al* 2010). In general, Papuan people consider that the forest is like a mother to them, and they rely on plants and wild animals from forest for source of food, clothes and shelters, and also for cultural purposes (Shaw 1969, Hyndman 1984, Kwapena 1984, Petoczs 1978 and Pattiselano 2003).

The knowledge of sociocultural aspects on *Goura victoria* in the northern Papua was not clearly known, particularly in all sites of this *Goura* study. It was found that the local hunters in Buare never use *Goura victoria*'s feathers or other body parts in traditional ceremonies, while in Bonggo area, the hunters only used the bird's feathers to decorate their hut in the forest, as a sign of their success in bird's catching. It was also found in Unurumguay and Supiori, where the feathers have never been used by local people in any traditional rituals.

Wildmeat play important role as the protein source for local people. Buare's hunters who live in Dabra, a small town in Mamberamo, usually sell fresh wildmeat from Buare's forest in market days. This situation was different from other three study sites, where hunters in Supiori, Unurumguay and Bonggo commonly sold their wildmeat door-to-door due to unavailability of market place. It is similar with the local people from other areas in New Guinea, where hunting has become the most important livelihood because it provides basic need on animal protein for the family and also money from wildmeat sale (Sada 2005, Mahuse 2006, Pattiselano and Mentansan 2010, Pangau-Adam and Noske 2010, Bulmer 1968, Sillitoe 2001, Cuthberth 2010). In this study, hunting on *Goura* might be considered as a part of traditional hunting of several tribes and clans in the northern Papua.

Hunting has become the part of local cultures and way of life, and already lasted from generations both in Papua and Papua New Guinea (Pattiselano and Mentansan 2010, Bulmer 1968, Dwyer 1974, Kwapena 1984). This similar situation occurs commonly in other regions in Indonesia, such as North Sulawesi, Kalimantan, Nusa Tenggara Timur, Java and Sumatera (Wadley *et al* 1997, Lee

2000, Farida *et al* 2001, Purnama and Indrawan 2010), as well as some areas in Asia (Kaul *et al* 2004, Rao *et al* 2005, Aiyadurai *et al* 2010), and in Madagascar (Randrianandrianina *et al* 2010) and Africa (Pailler 2009, Colell *et al* 1994).

Another tradition in local hunting system in Papua is that hunting determinedas the man-fully activity. Sometimes, hunter's family including children can join hunting if hunting sites are far from the village or if the hunting may take a long time activity. This study revealed that although the wifes may join their husbands during hunting time, they should stay around the huts. The hunter's wifes from villages of Buare and Unurumguay usually looked for their children, collected wild vegetables, fished in small ponds and creeks around their huts while waiting for their husband. Likewise, when following their husbands in the forest, women in some parts in Mamberamo region would run the harvesting sago and collecting wild vegetables in the forest as long as their husband hunt (Boissiére *et al* 2007 and Padmanaba *et al* 2012). In this area, only men including young men are considered to absolutely know about the forest and hunting boundaries between tribes and between villages. Another study also added that women in New Guinea were rarely or never getting involved in hunting or any related activities (Bulmer 1968). This study showed that basically the women already had enough household duties, so they do not need to go for hunting. This condition was similar with another study in New Guinea Highlands, where the women were never observed pacing through with hunting weapons and equipments or going into the forest for hunting practices (Silitoe 2001).

However, in some other places in the world, women are involved in wildlife hunting. For instance, women from Aka tribe in Central Africa are allowed to participate in hunting wild animals (Noss and Hewlett 2001), as well as the women from ethnic Agta in Philipines (Goodman *et al* 1985, Griffin and Griffin 2000), and Aborigines in Australia (Bird and Bird 2008). Similarly, women and their boys in Panama are practicing such a simple hunting specifically hunting on birds, that the women usually use traps to catch birds, while boys catch the small birds using slingshots (Smith 2005, 2010). In fact, whatever animals from boys' hunting, it still has no significant effect to what they eat, but it may have general impacts on the vulnerability of bird population and bird ecological function like the loss of pollinators (Smith 2005, 2010).

Recently there are some changes on women involvement in hunting by the Genyem tribe in Jayapura. The women in this area already begin to hunt wild animal specifically small understorey birds, phalangers and bandicoots. These animals are usually trapped using foot-snares that could be easily set up by women (Pangau-Adam, *pers.comm.*).

The age range for hunters in this *Goura* study is wide enough. Papuan hunters usually begin their practices on the age of 11 years and continue hunting until the age of 60 years old, as long as their physical condition can support them for hunting. In fact, around 45.02% from all respondents in this study were on the age of 25-49 years old. This condition is similar to hunters from Malinké ethnic group of Guinea-West Africa, that are usually on the age of 25-50 years old, have more than 10 years hunting experience and doing hunting as long as their physical health and ability can support them (Pailler 2009). Other study on hunting in Bioko Island-Africa showed other range of hunter age, around 14-72 years old, with productive range mostly around 20-35 years old (Collel *et al* 1994). Likewise, hunters in the highlands of Papua New Guinea also range between 20-60 years of age, and at the age of 50-59 years old as the most experienced group with higher level of hunted animals brought home than younger hunters (Sillitoe 2001). Slightly different is a study in Papua New Guinea that showed the younger-age group of hunters (Mack and West 2005). This study mentioned that a range of hunter age can be around 26-45 years old and it seems that they were more success in hunting than hunters from younger and older age classes. Dwyer (1974) stated that hunters ranging on the age of less than 30 years old can have higher frequency and more success in hunting than hunters on the age of more than 30 years old.

This study also revealed that hunters in all study sites mostly hunted alone and not in a team. This finding is supported by the hunting study amongst Genyem people in Papua (Pangau-Adam *et al* 2012). The main reason is that they have to share the hunting animals among the group member if they hunt in a group, so at the end each member will get less harvest compared to that of single hunting. However, they would go hunting in a team of two or three hunters, particularly when they need more meat for a big event or clan meeting. Similarly, the hunters from Rofaifo ethnic group in Papua New Guinea like to go hunting alone rather than in a team, although they sometimes doing that in a team (Dwyer 1974). The same hunting type also applied by the hunters from Agta ethnic group in Phillipine (Griffin and Griffin 2000), and   in Panama where the hunters usually go alone or in small groups (Smith 2010).

The hunters in all study sites prefer to hunt during the night, even though the bird is not nocturnal animal. They reported it is easier to find *G.victoria* when it sleeps on the tree, so they used flashlight or torch to make the bird daze. During the day, the hunters should prepare the hunting equipments and set up the traps when pacing their tracks. Correspondingly, several wild animals in Papua New Guinea are nocturnal animals, so hunters may catch both diurnal and nocturnal birds, reptiles or fishes (Dwyer 1974). Overnight hunting was found more popular on Agta ethnic group of Philippines because Agta hunters considered that the quantity and size of hunted animals at night were more and bigger than that obtained during the day (Griffin and Griffin 2000). The same reason was also reported by the hunters in Bioko Island in Africa (Fa 2000).

This *Goura* study also notified that hunters in all sites have no specific season on hunting. The hunters can go hunting at any time in a year both in rainy season and dry season, but some hunters reported that they may catch limited or less hunted animals during the rainy season. This situation have also described by the hunters among Genyem community in the northern part of Papua (Pangau-Adam and Noske 2010, Pangau-Adam *et al* 2012), and in other tropical region such as in Bioko Island-Africa (Fa 2000).

The hunters in this study prefer to go hunting in far forest areas, at least more than two km from the village. The main reason for that is related to the hunting target animals. The main target animals were wild boars and cassowaries, but not *G.victoria*.   If they plan to catch *G.victoria*, they will set up the foot-snares in the further area, because this bird can only be found in remote sites within the primary forest. To reach the far distance hunting grounds, hunters sometimes should join the logging workers in logging vehicles or log-hauling trucks or asked the owner of other vehicles to join in the truck to reach far distance hunting area. Other study showed that hunters in Panama frequently left bird as their main target when they go hunting far from their village, about more than two kilometers in distance and they only focused on large animals like mammals (Smith 2005, 2010). This also occurred in all study sites that the hunters typically focus on large understorey birds and high valuable species when they want to hunt birds. They should choose more distant area for hunting birds rather than when they hunt mammals, because the large forest dweller birds usually disappear from disturbed areas (Mena *et al* 2000, Smith 2005 and 2010). In general, the most reason for selecting distant hunting area is the high yield, since they discover more depletion on wild animals around their villages (Bodmer 1995, Fa 2000, Mena *et al* 2000, Pangau-Adam *et al* 2012) and in the forests close to the village.

5.3.2 Traditional knowledge of Papuan peoples in hunting

Papuan people already have a kind of traditional wisdom or knowledge regarding to hunting practices. This wisdom consists of some important rules controlling some taboos, animal species, specific forest areas and also indigenous cultural rules in hunting (Kwapena 1984; Pattiselano 2006 and 2008, Pattiselanno and Mentansan 2010). This *Goura* study could collect from all sites six important traditional rules related to the hunting practice. The first rule is an obligation for all hunters to get hunting permission from tribe's chef (Ondoafi) who also sometimes acted as village chief for such administration purposes. This obligation is usually followed by the instructions of specific permitted areas for hunting. It means the hunters should avoid some forest areas that customary established as ancestral lands, and they must obey the rule, particularly if the hunters come from other villages.

The rule must be concerned to avoid any possible undesirable things, e.g.: conflict between clans, as had occurred in Buare village. If a hunter break the rule, he should get punished and pay a fine or a sum of money. Study in Sorong

Selatan Regency reported that certain forest is a sacred place specialized for worship of the tribal ancestors, so the hunters are prohibited to practice hunting around or disturb that area (Pattiselanno and Mentansan 2010). This also occurs in India, where certain forest areas are established as sacred places, and the citizens are prohibited to perform any activity (Madhusudan and Karanth 2000). The violation of this rule can result on an accident or any supernatural sanctions for the hunters (Cinner 2007). The similar rule also applied in Bioko Island-Africa, where the people have taboo on hunting and eating some particular animals (Collel *et al* 1994). In Papua, even if the animals are unintentionally trapped or captured, it is believed that they will bring bad lucks for the hunters.

The presence of customary sacred forest, specific rules and traditional rules on wildlife hunting also exist in local people in West Kalimantan (Wadley and Colfer 2004). Likewise, Mamberamo community strongly believes on sacred area in their forest, for instance local people will never take any kind of forest products from the area of Foja Mountain (Padmanaba *et al* 2012). The community considers that the violation of this rule will lead to the illness into the death or impact a natural disaster like lightning, storm and heavy rains in the entire village. This belief also recorded from Arfak community in Manokwari, West Papua (Makabori 2005).

The second rule in traditional hunting is concerning the women and family involvement in hunting practices. The hunters usually refuse to bring women with them during hunting time basically due to the physical condition of women. Hunting needs excellent physical conditions of hunters, so the women are considered to be unable for this activity and should be in her right place or stay at home (Bulmer 1968 and Sillitoe 2001). In fact, the traditional rule demands that hunters in Papua should not go for hunting, if his wife is pregnant. The violations of the rule usually result on fruitless hunting. The hunters also believe that their babies will be harmed or born flawed if they still insist to go hunting during their wife pregnancy. It is like a natural punishment for them because the fathers wanted to kill the animals or even killed them during hunting. Except the hunter's personal reasons, there are no specific and scientific explanations for this "hunting-pregnant wife" problem (Lambek 1992). It is assumed that the reason is much related with women nature, common ancestor's belief on women's role in regulating all aspects of life including in heredity inheritance (Cinner 2007, Pattiselanno and Mentansan 2010).

Regarding to legal restriction on hunting, *G.victoria* is already protected under some legal acts. This bird is protected under the Act of Republic Indonesia No. 5/1990, for Conservation of Natural Resources. It is followed by the Decree of Agricultureal Minister of Indonesia No. 301/1991, then by Law Act No.7/1999 concerning Plant and Animal Preservation, and also Law Act No.8/1999 about Wildlife Utilization. Long before those legal acts, *G.victoria* has already been protected under the Decree of President of Republic Indonesia in Law Act No. 43/1978 that ratified CITES (Appendix II) and the EC-CITES regulation (JNCC 2005). This bird is listed as vulnerable species in the IUCN-Redlist (IUCN 2012). Therefore hunting and trading of *Goura* is not allowed. In Papua New Guinea, there are similar regulations of prohibiting the hunting and trading on Genus *Goura* (Shaw 1969).

This study showed the low implementation of law enforcement on hunting, specifically in relation with this species. There were only 26.5% from all respondents whose avoid to hunt the bird due to protection law basis. As reported by the respondents, they have no sufficient information about the related laws or they do not understand the legal acts. Lack information on protected species and hunting prohibition might be the main reason of low number of hunters that keep away from hunting on *Goura*. About 33.11% of hunters did not know that *Goura victoria* is protected bird, so they keep hunting on *G.victoria*. In other part of Papua, beside northern cassowary and other bird species, *G.victoria* is also hunted for sale (Sada 2005, Mahuse 2006, Suryadi *et al* 2007, Pangau-Adam and Noske 2010, Pangau-Adam *et al* 2012). *Goura* spp are trading not only in Papua but also in national and international markets. Massive trade of all species of *Goura* has been reported since 1997 as well as the illegal export of *G.victoria* to overseas countries (King and Nijboer 1997) and to the Philippines in 2010 (Profauna 2010). As a comparison, people of Papua New Guinea already understand about hunting restriction on certain bird species, e.g. the Lesser Bird of Paradise (*Paradisea minor*). However, the facts remain that the bird is still hunted for its feathers, which can be used as emblem and complement in customary or traditional ceremonies (Shaw 1969, Kwapena 1984).

5.3.3 Hunting attributes

The hunters in all study sites preferred to use foot snares rather than using other equipments. These equipments along with machetes, knives and flashlight are also used by other hunters in Indonesia, (Pattiselano dan Mentasan 2010). The use of foot snares can be more fruitful and the possibility to catch live *Goura* is higher than using other equipments. Because, the living birds have high prices in the markets, the hunters should select the proper equipments for capturing them alive. They also hope to catch adult birds with their chicks, though such chance rarely occurred.

Hunters in Papua usually make foot snares manually from nylons or other natural materials such as stems of lianas that looks like cords and tied them up to be some ropes. These ropes were generally set up and distributed in 10-30 pieces at each playing ground, feeding ground, water pools, and in the forest floor, especially in the area that full-grown by *Ficus* spp. During fruiting season, the fruits of *Ficus* spp. usually emerge from the whole part of the stems above ground. These fruits known as favourite fruits for *G.victoria*, so the hunters will set up the traps on the ground around the tree.

Furthermore, there are several hunting strategies and their modifications in New Guinea (Bulmer 1968). One of the strategies is "ambush" with its modifications, which let the hunters waiting passively while hunting or capturing the animals. When catching big *Goura* birds, the hunters will wait the birds near their feeding trees, or set up the traps around their playing grounds or water pools, sometimes climb up the trees to catch the birds in their nests, or check the nest then take away the eggs from it. Some hunters from all study sites also reported that *Goura* can be catched by finding its roosting tree at night using flashlight or torch (fire), because the light will attract the birds. This light-attraction method becomes a kind of modification from 'ambush' technique. The hunters stated that this method can result in capturing more than a bird at once hunting trip, but this is more applied to the high-experienced hunter who know better the position of sleeping trees. It is also reported that high experiences commonly influence the hunter's ability for catching *G.victoria* at its feeding trees, playing ground, nesting trees or at its roosting trees (Bulmer 1968, Dwyer and Minnegal 1991).

Generally, hunters in New Guinea are mainly hunting and catching live animals, raising up before consuming them (Bulmer 1968) though actually they have no particular system on raise captured animals from the forest (Pangau-Adam and Noske 2010, Pattiselanno and Mentasan 2010), Mahuse 2006, Torobi 2005, Sada 2005). However, this practice changed through the time and recently many local people begin to raising up the captured *Goura* and wild boars for consumption or sale purposes. Some birds such as cassowaries (*Casuarius* sp.), crowned-pigeon (*G.victoria*), cockatoos (*Cacatua gallerita*) and lorikeets (*Lorius lorius*) are also captured and kept by hunter families for certain time before being sold (*pers.obsv*, Pangau-Adam, *pers.comm*).

About 100% of hunters in all study sites prefer to catch or trap *Goura* alive. The hunters informed that when they used foot snares, they should check the traps every day. This is important, because when a *Goura* is trapped and left in the forest, the bird can die due to rotten wounds in its foot, or may be eaten by other animals like wildboars or monitor-lizards (*Varanus* sp). This information is confirmed by the study on hunting patterns in Genyem, Papua and in Panama as well, reported that hunters regularly check the traps to prevent the captured animals being preyed by predators (Pangau-Adam and Noske 2010, Smith 2005, 2010).

Furthermore, the *Goura* study revealed that the quantity of using air-gun for hunting in all study sites is still limited. It was only about 3-22% or in average 11.3% of the hunters who admitted having air-gun. If they own this weapon, it was mainly used to shot paradise birds and fruit doves, and not specifically targeted on *G.victoria*. This bird is very sensitive to any sounds, noises, or disruptions, and never perch longer in certain branch of tree as well.

The limited use of air gun in hunting of Papuan people commonly occurs due to their low economical conditions. The hunters cannot afford to buy the costly weapon cartridges, and they have limited skills to maintain even if they have it (Pangau-Adam *et al* 2012). Therefore, hunters in all study sites mostly use traditional equipments and methods. Some area in the northern Papua like Genyem and Nimbokrang (Pangau-Adam and Noske 2010, Mahuse 2006), Nabire (Pattiselano 2007), and Bonggo as one of the study sites have resettlements of transmigrant people from Java. These are located among the settlements of local people. The communal life, routines, knowledge and experience have been exchanged between the transmigrants and local peole, and resulting on many changes in local people life style. Recently, some local people are already skilled in using air-gun to hunt paradise birds or fruit doves, although they also use such borrowed weapons.

There is report of respondents to use a high-calibered weapon. This weapon only limited to police or soldiers, when they hunt wild deers or wildboars occasionally, and was also used when the local people need wild meat in big quantities for customary or religious ceremonies (Pattiselanno 2006, Pattiselanno and Mentansan 2010). On the contrary, the use of air gun during hunting is a common method among hunters in Papua New Guinea (Kwapena 1984). Similarly it is also showed by the Iban tribe in Kalimantan, who are rapidly learning how to use gun for hunting (Wadley *et al* 1997). If used guns increase among the Papuan people, this may heavily threaten the wildlife animals. The uses of riffles and shot guns have also become a common method among hunters in Panama, and might kill about 49% of their hunted animals (Smith 2005, 2010), and hunters from Agta ethnic has killed about 44% of their hunted big animals by using gun (Griffin and Griffin 2000).

In Papua, dogs are usually used to hunt deers and wildboars, but rarely for cassowaries, and small animals (Ariantiningsih 2000, Cahya 2000, Andoy 2002). This study showed that only very few hunters (11.92%) used dogs to hunt *Goura*. The disadvantages of using dogs for hunting *Goura* are: the bird will be torn and swallowed directly, or it will fly instantaneously due to being startled. Therefore, most hunter choosed to use foot snares rather than the dogs in hunting *Goura*. In the same way, Rofaifo people in Papua New Guinea use dogs to help them in hunting both alone or in group (Dwyer 1974). They do this although sometimes the dog will disrupt the preys. Basically, dogs are the main aids to hunt and kill the hunted animals in Papua New Guinea (Mack and West 2005). It is also reported that most of hunted animals in PNG were captured and killed by the dogs (Mack and West 2005). Generally, decision to use dogs for hunting in New Guinea is based on hunting strategy and animal targets (Bulmer 1968).

The hunting activities in all study sites were relatively high compared to that in other areas of Papua, e.g. hunting in Waropen (Sada, 2005). About 83% of hunters were hunting at least twice a week, because they want to get as much meat as they can. The meat was needed as protein source (*wildmeat*) and for sale to get cash money, when there is an excess meat of their consumption (*bushmeat*). This is comparable to the cases in PNG (Mack and West 2005). High frequency of hunting activity also occurred, because there is a responsibility to check the traps regularly, or at least once in three days. In addition, hunters had no particular time or day, or particular season during a year for practicing hunting. They usually go whenever they want or they need to hunt. There is no specific schedule on hunting season, because the hunters assumed that hunted animals are abundantly available in the forest.

Moreover, Papuan hunters never focus on certain animal species, when they go hunting. This study listed at least 15 species of hunted animals in all study sites including *Goura victoria* (table 4.7). This list was very short if compared to a list of 135 species (696 individual) of hunted animals around CMWMA-Papua New Guinea (Mack and West 2005). Local people in that location hunt wild animals regularly, which including 264 birds from 86 species. Similar practices were carried out in many places around the world. For example people from Maya-Mexico ethnic group could hunt about 584 species of animals and 34% of them were birds (Jorgenson 1995). The local people in my study sites are not fully dependent on hunting activities to fulfil their protein need. For example in Supiori, most hunters were also go for fishing, and therefore only 12 animal species were recorded as the hunted animals.

In one-month hunting time, people from Moka Bubis ethnic group from Bioko-Africa can hunt about 332 animals of 40 species including four bird species (Colell *et al* 1994). People in north-eastern Madagascar also reported that they captured 23 species of Mammals (Golden 2009), and the hunters from Iban tribe West Kalimantan-Indonesia could hunt about 34 species of primates in the same range of time (Wadley *et al* 1997). It seems that the hunters in all study sites have preference of target animals and focus on the large and medium sized animal species like wild boar, rusa deer, *Goura*, cassowary, crocodiles and bandicoots.

Other study in Papua, reported the similar number of hunted animals; 18 species (Pangau-Adam *et al* 2012), while seven species of hunted wild animals were listed from Maybrat ethnic group in Sorong and only six species were listed from Waropen (Pattiselanno and Mentansan 2010, Sada 2005). Hunting system in Papua can be classified as a non-selective system. But in the study area, the hunters are not targeting *Goura* as the main animal species to be hunted. In Mamberamo area, other wild animals like wildboars, cangaroos, cassowaries, cockatoos and lorikeets were more freguently hunted and sold rather than *Goura victoria* (Boissiérre *et al* 2007, Padmanaba *et al* 2012).

5.3.4 The utilization of *Goura victoria*

As in other tropical regions, wildlife hunting and trading of captured animals are practiced for the nutritional needs and economical purposes (Milner-Gulland *et al* 2003). Hunters in all study sites attempted to capture *Goura* alive, because there was a good demand from villagers and people living in the town to buy and rear this bird as a pet. The price of an individual bird could be around US\$ 1 to US\$ 100. In the Genyem community this bird was found to be priced around US\$ 50 (*pers. obsv.*). In all study sites, the price was rather low, which ranged around US\$5 to US\$ 15 per each bird. There was no standard price for selling *Goura*, so the hunters might determine the price by themselves. Sometimes, if the hunters really need cash money, they sell the bird in a low price, because the buyers usually handle the prices.

In fact, the quota on hunting and trading *G.victoria* in Papua is determined as zero quotas because this bird is listed as protected species. In the reality, trading on *Goura* is increasing as shown by the local hunters that tend to catch and sale the living birds. The highest price of *Goura* might reach US\$ 200 for each bird (Suryadi *et al* 2007), so this has become the main reason of trading *Goura* in regional, national and international level. Morever, the selling price of *Goura's* meat was cheaper than that of the live bird. In Unurumguay the price of a live *Goura* was valued for about US$ 10, while roasted meat of *Goura* only priced around US$ 1.5 - US$ 3. When captured alive, the hunters usually bring the bird to the market for sale, or if there is no market in the village they just offer it from door to door. In all sites, *Goura*'s meat is rarely sold outside the villages, due to the limited transportation system and the absence of nearby markets. Sometimes people from the town visit the villages to find and buy live *Goura* and kept the bird as pet. Other parts of *Goura* like feathers and bones were not exploited by the hunters.

This study showed a limited consumption of *Goura*'s meat in all study sites. About 24.5% of hunters and their family ate the meat of *Goura* occasionally and 75.5% of all respondents rarely consumed the *Goura* meat. *Goura* meat was rarely or occasionally consumed, because the hunters choose to sale it and replaced their protein source with other animal meat, for instance fish. The hunters usually sell most of all meats (fresh or roasted) from their hunted animals including *Goura*'s meat, and only consume little parts left. This also reported by the study in Genyem community, that hunters typically remove specific parts of hunted animals such as head, legs and intestines (±1-5 kg of each animal) to be consumed by the family, and take the meat to sell (Pangau-Adam *et al* 2012).

People in Mamberamo area considered *G.victoria* not only as hunting target for cash income but also as food sources. If they catch other large birds and *Goura*, the meat of *Goura* will be consumed, and the meat of other birds will be sold freshly or smoked. The price of meat of others birds such as cassowary is more expensive than *Goura*'s meat. Additionally, the chicks of cassowary, live cockatoos, and lorikeets were also more expensive than *Goura*'s meat, because all of those birds can be easily raised as pets and then sold, if the birds reached certain size.

In Supiori, hunters and their family do not heavily rely on bushmeat or wildmeat because, they can go fishing to get fish and shrimps for their protein needs, or buy those protein sources from their families who work as fishermen.

According to some anthropological studies, the daily real life of Papuan people in the study sites much relies on the forest and natural resources around them (Pattiselanno and Mentasan 2010, Boissiére *et al* 2004 and 2007, Mansoben 2005, Apomfires 2002, Lamera and Siregar 1992, Sanggenafa 1992, Dumatubun and Wanane 1992, Apomfires and Sapulete 1992).

5.4 The impact of traditional hunting on the population of *Goura victoria*

It is considered that the main pressure for natural resources including wildlife and forest are human activities for fulfilling their daily needs. Nowadays, the utilization of forest and natural resources occurs both for subsistence and economic purposes (Ellis *et al* 2012, Morris 2010). For instance, local people harvest non-timber forest products (Chamberlain *et al* 2004, Kuster *et al* 2006), extract minerals and open mining in the forest area (Philips 2001, Miranda *et al* 2003, WWF Global 2012), and create big plantations (Gillison *et al* 2004, Danielsen *et al* 2008, Yaap *et al* 2010 and Obidzinki 2012). They also establish farms and agricultural lands (Eden 1993, Peroni and Hanazaki 2002), harvest timber through logging activities (Wilson and Johns 1982, John 1983[a], 1985, Johns and Johns 1995, O'Brien *et al* 1998) and of course practicing hunting (Colell *et al* 1994, Jorgenson 1995, Alvard 2000, O'Brien *et al* 1998, Kaul *et al* 2004, Fa and Brown 2009, Aiyadurai *et al* 2010). Each activity mentioned above will influence the existence and sustainable of wild plants and wild animals in the forest.

The Government of Indonesia has already ratified the regulation concerning hunting on wildlife so-called Law Act No. 13/1994 on Hunting and Hunting Animals. In this regulation imply that wild animals as protected animals should not be hunted, captured, and traded. Consequently, anyone who breaks this regulation will go under sanctions and get punishments. However hunting practices on protected animals still commonly occurr in Papua including hunting in protected areas. This study encountered that hunting on *Goura* was also undertaken in the Nature Reserve area in Supiori. This practice was carried out by both local residents and the people from outside of Supiori region, who work in construction projects of Supiori Government program. In all study sites, hunting on *Goura* was considered as illegal hunting. Furthermore, illegal trades of wild animals including *Goura* also occurred in several areas in the northern Papua (Padmanaba *et al* 2012, Pangau-Adam and Noske 2010, Suryadi *et al* 2007, Boissiére *et al* 2004 and 2007, Mahuse 2006, Sada 2005). Other cases of illegal hunting and trade on the wild animals also occurred in Lorentz National Park and Wasur National Park, Papua (Cahya 2000 and Andoy 2002, Suprayitno 2007).

The cases of illegal hunting and wildlife trading also happened in other protected areas in Indonesia as well. For instance, there were illegal hunting and trading of Seram cockatoo in Manusela National Park and Mount of Sahuai Nature Reserve in Seram, Mollucas (Kinnaird *et al* 2003), followed by illegal hunting on Sulawesi crested-black macaques in Mount Sibela Nature Reserve in Bacan, North Mollucas and in Tangkoko Duasudara Nature Reserve in North Sulawesi (Rosenbaum *et al* 1998). Another case of illegal hunting and trading also occurred on the Bornean Peacock-Pheasant in Bukit Raya National Park, Central Kalimantan (O'Brien *et al* 1998). Those illegal practices were also found in other place in the world, such as in Africa (Muchaal and Nganjuh 1999), Papua New Guinea (Mack and West 2005), Myanmar (Rao *et al* 2005) and in Madagascar (Garcia and Goodman 2003).

The report on bird hunting in Asia is absolutely limited (Corlett 2007, 2009, Aiyadurai *et al* 2010), although many studies had mentioned that wild animals were hunted in many places in this continent. In Indonesia, large birds like cassowaries, hornbills and megapods were hunted for their meat and eggs (Padmanaba *et al* 2012, Pangau-Adam and Noske 2010, Pattiselanno and Mentansan 2010, Mahuse 2006, Sada 2005, Johnsons 2004, Agerloo and Dekker 1996, O'Brien and Kinnaird 1996). Other bird species like grouses, partridges, pheasants and the Galliformes group were also hunted in several Asian countries (Aiyadurai *et al* 2010, Brickle *et al* 2008, Kaul *et al* 2004, Keane *et al* 2005, McGowan and Garson 2002, O'Brien and Kinnaird 1996). In New Guinea, hunted birds include several species of paradise birds and crowned pigeons (Healey 1978, Kwapena 1984, King and Nijboer 1994, Buiney 2006, Pangau-Adam and Noske 2010), as well as some species of cockatoos and lorikeets (Walker *et al* 2005, Marsden *et al* 2001 and Kinnaird *et al* 2003,). All of these birds are hunted for different purposes including food, medicine, and pets, and in some cases they are hunted because the birds are considered as pest animals.

Large birds and many terrestrial birds especially forest-floor dwellers are particularly more vulnerable as hunting targets compared to small birds (Pangau-Adam and Noske 2010, Haugaasen and Peres 2008, Smith 2005 and 2010, Peres 2000, O'Brien *et al* 1998, Bodmer 1995). Large birds generally have smaller clutch size, slow-motioned adults and have a low rate of reproduction (Smith 2005 and 2010), so hunting pressure may affect the persistence of those birds in the forest. Large birds usually become hunting target when big mammals are difficult to find, because these birds can have high-valued rather than other forest product in tropical forest area (Thiollay 2005, Peres 2000). Actually *Goura* can be classified as the large birds, but in fact its meat is not much for selling, and therefore the hunters mainly insisted to capture the birds alive. It is reported that *G.victoria* can fulfill 'the requirements' as hunted-animal target in several regions of Papua (Gibbs *et al* 2001, Baptista *et al* 1997, Beehler *et al* 1986).

This study showed that hunting activities have negative effects on the population on *G.victoria*. The results showed that the estimation of annual hunted on *Goura* in all study sites were around 83.20 -811.20 individual birds harvested in one year, both for self-consumption and sale. This result is highest compared to the hunting study in Genyem and Kemtukgresi, Jayapura Regency, Papua, that recorded 42-45 captured *Goura* in one year (Pangau-Adam and Noske 2010). Study on Cracidae in Peruvian Amazon documented unsustainable hunting practice on four species of large-bodied birds. In this study the reason of high number of *Goura* captured might due to the high frequency of hunting activity by the hunters, the increased number of hunters and easy access into the forest areas.

As the comparison, some studies on large birds in South American found the maximum sustainable yield-values for Crested Guan (*Penelope purparascens*) was 0.14-1.31 birds each km^2 (Smith 2010: based on study of Begazo and Bodmer 1998, Ohl-Schacherer *et al* 2007), and for the Great Currasow (*Crax rubra*) was between 0.16-0.26 birds per km^2 (Smith 2010: based on study of Begazo and Bodmer 1998, Silva and Strahl 1991, Ohl-Schacherer *et al* 2007), several studies also stated that the great Curassow will become the first disappearing species from its habitat because of hunting (Escamilla *et al* 2000, Peres 2000, Thiolay 2005 and Barrio 2011). For the Great Tinamou (*Tinamus major*) the estimation of annual harvest was around 0.26 birds per km^2, and the maximum sustainable harvest estimation value was 7.81 birds per km^2 (Gonzales 2004 *in* Smith 2005). The maximum sustainable harvest value of the Great Tinamou was higher than those of the *Goura victoria* (in this study). This might occur, because the great Tinamou is a highly secretive, cryptic bird and difficult to detect, although it has been reported as overhunted bird elsewhere (Thiollay 2005, Perrin 2009). Those studies and this current assessment in Papua can show the unsustainable practices of hunting on the large bird including *Goura victoria*. Unsustainable hunting may adversely affect the population abundance of the birds.

Eventhough people in northern Papua area and Supiori Island only used their traditional weapons and did not mainly targeting *G.victoria* but the hunting practices have negative impacts on the population of *Goura*. Wild animals such as *G.victoria* can be classified as long-lived species, with low rates of increase (low r_{max}), and long-generation time (Bodmer *et al* 1997). This species might be more vulnerable to the extinction than short-lived species that have high rates of increase (r_{max}) and shorter generation time.

Several other studies showed that local hunting as a sustainable practice is difficult to be confirmed (Barrio 2011, Begazo and Bodmer 1998, Ohl-Schacherer *et al* 2007, Franzen 2006, Noss 2000, Lee 2000, Fa 2000, O'Brien and Kinnaird 2000). It still needs more intensive and long time studies in other areas of Papua with supports of sufficient data for the comparison of hunting on *Goura* in Papua. These further studies are needed because some hunting activities already lead to the population decrease and even the local extinction of *Goura*. Increased hunting on *Goura* might be influenced by some other human activities such as forest conversion to oil-palm plantation and logging activity. The loggers used to hunt animals to fulfill the bushmeat demand for logging camps.

As in other tropical regions, wildlife hunting and trading of captured animals are practiced for the nutritional needs and economical purposes (Milner-

Gulland *et al* 2003). This study also revealed the change of hunting system in all study sites from pure subsistence hunting to a mixture of subsistence and commercial practices. Currently hunting on *Goura* is not merely for family consumption but also for commercial use, because this bird is a target of illegal hunting and illegal trading. These changes could increase the pressures on *Goura* population in its habitat. This current study may reveal the impact of hunting on the population of the bird. Eventhough it was only covering the four study sites areas in Papua, this assessment could make a significant contribution on the basic information of population abundance of *Goura* in Papua and the sustainability of bird hunting.

5.5 Participation of local community in *Goura victoria* conservation on indigenous knowledge basis

In general, hunting practice is a part of human activities that has been undertaken for so long and will always be carried out as long as hunting instinct and hunted animals available (Supriatna 2008). Hunting practices cannot be stopped or eliminated instantaneously, because this activity already became tradition for local people.

Historically, there is a prediction that hunting activity in Papua was practiced since about 35,000-40,000 years ago (Hope 2007). There are also some archaeological evidences that support the idea, where the hunting tradition has been undertaking from the past until recently. Another assumption stated that wildlife hunters might have been living in Papua since about 46,000 years ago (Roberts *et al* 2001).

Basically, local people in Papua can be divided into four ecological zones related to their way to use forest resources including animals and plants within the forest (Mansoben 2005). The zones basically influence the patterns of social-economical adaptation of Papuan people. Every group of local community in Papua will have particular patterns of life and activity, such as in livelihood, social dan cultural aspects, associated with their specific zones (Mansoben 2005). People who live in the zone of swampy areas, coastal and riverine areas, also in the zone of coastal lowland will practice hunting as an alternative and additional activity. It is quite different in the zone of hills and small valleys where the people depend heavily on hunting strategy and practices. People in highlands zone are more specific than in other zones. They rely on farming and rising pigs, and hunting is done as a complementary activity.

Furthermore, hunting practice is known as traditional activity for local people. This activity cannot be separated from their daily life and has become their major livelihood. In several regions, hunting practice is particularly intended to fulfill need of animal proteins, non-profitable purpose, and to maintain hunter's pride. This study showed that merely 20% of wild animals captured are not protected by Indonesian law and the rest (80%) are under protection of both Indonesian and international regulations. This condition also occurs to *G.victoria* that has been hunted though the bird is protected by laws.

In general, the threats on biodiversity have many characteristics globally. The threats for biodiversity can come from top-down or bottom-up. The threats can also be classified as global threat and landscape threat, or direct threat versus social and institutional threat, and also natural threat and human threat (Birdlife International 2001). Direct threats usually appear from social problems like pressure from human population, consumption on natural resources, poverty and unequal access to natural resources (Kaninen *et al* 2009, Frazier 2007). Facing many threats, protection on natural resources needs the holistic approach involving all stakeholders including local communities.

The involvement of local people can play important role for forest and biodiversity conservation. The development in technology standard and capacity in all aspect of human life hold important role in changes of people knowledge and implementation of conservation values. For example, Arfak ethnic in Papua has such a traditional concept of natural conservation (Makabori 2005). The Arfak people regard their natural environment and forests as *"Igya ser hanjop"*- Arfak phrase for "(our) forest should be secured". Unfortunately, this concept tends to be violated in the last decade by its young people. This new generation considers that the concept might only restrict their way to use natural resources, such as wildlife hunting. However, the old generation still respects the concept and performs the customary rituals actively.

Although the sanctions and punishments under some legal Acts and Government Regulations were already established, there is no clear and strong law enforcement in hunting problems (Saleh 2005, Hardjanti 2005). In Papua the local legal regulations or Specific Regional Regulation (PERDASUS – Peraturan Daerah Khusus) for controlling and protecting the Papuan unique and valuable wildlife have not been established yet. Although until recently, Papua and West Papua have already been taking a decade of Regional Autonomy Government system, there is still a lack of such legal acts on protection and concervation endemic wildlife in Papua from illegal hunting and trading.

Local people basically want to be involved in any conservation actions on *G.victoria* and other wildlife species using their traditional wisdom and customary knowledge. This study revealed that the hunters had no information on protection status of *Goura*, and they have never been asked to involve in any socialization event about wildlife protection in Papua. This condition can support the conservation activists to get more understand of all customary rules and traditions in each tribe or clan in Papua, and then learn their traditional knowledge and wisdom about ecological aspects and natural resources (Pattiselano and Arobaya 2013). The conservation activists should build a good relationship and trust with local people to make the learning and accepting process on conservation easier (Boissiére *et al* 2004). It is better if these processes can be undertaken in simple way, with local language, related to local values and cultures, and also matched with the rational level of local people.

The efforts to learn traditional knowledge from particular ethnic groups or tribes have already suggested some years ago (van Vlieth and Nasi 2008). As the tribes in all study sites have traditional wisdom on using forest and natural resources, there is possibility to improve and use that for conservation purposes. These conservation efforts should combine with such ethno-biological approaches to find out social and cultural factors which influence hunting activities, e.g. the case with local people in Gabon, Africa (van Vlieth and Nasi 2008). The results can be used to overcome the hunting problems, specifically hunting on protected wildlife. More detailed idea that using social approach (including approach of anthropology, ethnology and social-economy) together with natural approaches (considering ecology, botany, pedology and geography) can figure out the condition of local people (Sheil *et al* 2004). The knowledge and all solutions for people daily life related to the use of natural resources, including hunting will be known through those approaches.

In facts, the approach of *Community Based Nature Resource Management* (CBNRM) has been established in Papua in the last few years. For example, this approach has been carried out by the PtPPMA (Limited Association for Assessment and Empowerment of Indigenous People) in customary land of Nambloung and Kemtuk in Jayapura Regency, Arso in Keerom Regency and Knasaimos in Sorong Regency (Wamebu 1999). Similar action has also been implemented in Mamberamo region. In this area, Conservation International (CI) worked together with CIFOR to involve local people in managing their own natural resources (Boissiére *et al* 2007, Padmanaba *et al* 2012). These NGOs used MLA (Multidiciplinary Landscape Assessment) to identify all important natural resources for local communities within forest landscape. Also, the comparable method and approach have been used by management of Wasur National Park Merauke and WWF Papuan Region to manage natural resources in the Park (Supriatna 2008). Similarly, the involvement of local people has been developed by the leaders of Teluk Cenderawasih (Cenderawasih Bay) National Park through mapping action and zonation system based on their local and traditional values or norms to use their marine products and commodities (Fatem *et al* 2011).

The main focus of CBRNRM and MLA system in Papua is supporting local people in each village to participatively map their own natural resources. There is a potential, that the local people can be able to record all of their hunting areas, sago-palm farms, villages, sacred places and customary lands, or any taboos within their community. These two methods could be carried out by trainings on capacity building, like some collectively actions between CIFOR, CI Papua Program, LIPI and local people from the villages of Papasena I and Kwerba in Mamberamo (Boissiére *et al* 2004, 2007 and Padmanaba *et al* 2012). Basically, the trainings should involve multidisciplinary stakeholders and local people from Papua as natural resources'owner.

These trainings and mapping actions through CBRNRM and MLA systems can allow local people to understand clearly all "do's and don'ts" in taking their daily way of life, based on their own traditional rules (Boissiére *et al* 2004, 2007 and Padmanaba *et al* 2012). This knowledge and rules can be used to prevent any negative pressures from outside, specifically in regard to natural resources use. The sanctions when these rules are violated also need to be provided and determined before more legal processes undertaken.

In general, it is expected that CBRNM with its modifications, simultaneous with lasting and intensive guidances from multistakeholders can enlighten, motivate and change the thinking manners of local people (Boissiére *et al* 2004, 2007 and Padmanaba *et al* 2012). Afterwards, the people and particularly the hunters can accept and imply conservation concepts in their daily life through socialization on hunting rules and education on conservation actions.

There is another fact on supernatural beliefs that applied extensively among traditional and local people in Papua (Mansoben 2005). These beliefs also include all norms and value controlling human activities to utilize the ecosystems, for examples many ethnics belief their early ancestors were particular animals or tree (Mansoben, loc cit). This kind of norms and beliefs will prevent them from killing those particular animals, felling-down particular trees or destroy particular sacred forest, because they consider those actions will lead to vanish their own ethnic. Furthermore, these local people have their own social institutions and related instruments made by local and traditional community to manage the use of their forest and natural resources (Mansoben, loc cit). They adopt both of them as base for their daily life economically, socially and ecologically.

Ecologically, social instruments and institutions in local people emphasized that habitats of animals and plants are not the same (Mansoben, loc cit). Therefore there is a need to manage the use of habitats and wildlife in sustainable manner. One kind of the related management technique is a set of restrictions in harvesting or exploitation of products from forests or sea for particular time range

(Mansoben, loc cit). The restrictions are directed to give chance for restricted animals and plants to reproduce without being disturbed until proper harvest season. These restriction systems are implemented widely in Papua. For instance, there is '**Takayeti-Tiyaitiki** in Tablanusu and Depapre in Jayapura, also '**Sasisen**' in Cenderawasih Bay, '**Rajaha**' in Salawati, Sorong and '**Samsom**' in Raja Ampat (Mansoben, loccit). Another restriction and rule are also implemented among Sentani ethnic who live in the southern part of Cyclop Mountains, Jayapura. In this ethnic, traditional institution has specific management system so-called '**aniyo-erayo**'- an officer who organizes and controls the use of forest products, and '**yayo**'- who is in charge with management on wildlife hunting (Mansoben, loc cit)

The information mentioned above shows that the unsustainable use of natural resources can influence the forest-dweller communities, forest owner, and natural resources. In order to overcome these problems, it is necessary to consider more bottom-up approaches as conservation techniques (Tien *et al* 2009, Sodhi *et al* 2011[a] and Sodhi *et al* 2011[b]). Local communities should be involved in using, managing and protecting wildlife and its habitat.

5.6. Conservation status of *Goura victoria*

5.6.1 The role of habitat structure

This study showed that *Goura* was very rarely or never found in any opened forests, high disturbed forests or forest edges. These areas were usually high-populated by people and therefore more hunting and farming activities occurred. These findings were supported by the fact that *G.victoria* in northern area of Papua New Guinea was mostly encountered in the very remote forest due to hunting pressures around the villages (Coates 1985, Peckover and Filewood 1976). Similar situation occurred in Manokwari-West Papua, where the *G.cristata* is more difficult to find in secondary forest and exploited forest than in primary and remote forest (Kilmaskossu 2001).

Observation on *G.victoria* showed that many nesting trees can be found on the banks of small streams and flooded area in the forests that were located far from the villages. These areas are also ideal for feeding ground because they are rarely visited or walked through by local people (Beehler *et al* 1986). Furthermore, the lack of studies about habitat or other specific studies on ecological aspects required by *G.victoria* or other members of Genus *Goura* results on the insufficiently habitat information. This study assessed the general habitat structure in different study sites, where *Goura* could still be found. It was found that population density of *G.victoria* can reflect the forest condition and possible threats in each study site. The estimation of population density of *G.victoria* in Buare, Mamberamo was about 42 birds per km^2. This high density could be possible, because the forest area in Buare was mostly covered by primary forest and has no significant disturbance. In Supiori Island, the density estimation of *G.victoria* was about 40 birds per km^2, and this seems not much different from *Goura* population in Buare. Actually this area was established as a nature reserve and animal sanctuary, but the lowland forest was already disturbed through a variety of human activity in utilization of the natural resources. The other study sites showed the population density of *Goura* to be about 31 birds per km^2 in Unurumguay and 13 birds per km^2 in Bonggo. Both areas represented the most-disturbed areas that have been cleared for logging, plantations and resettlements. In addition, these areas are not categorized as protected area, therefore the use of forests was found more intensive compared to Supiori site.

This study might reveal that the population density of *Goura* tends to be high in intact forests with less or without human activities. *Goura victoria* is a primary forest-specialist, which much depends on the existence of primary forest hence their main habitat should be protected.

5.6.2 Conservation assessment of *Goura victoria*

Indonesia has become well-known for having the eighth-largest natural tropical forest in the world (FWI/GFW 2001). Unfortunatelly, this country already had forest degradation rate about 1.87 million hectares each year in 2000-2005 and it is reported that forest cover in Indonesia tends to decrease through times (FWI/GFW 2001). Another information stated that forest degradation rate in Indonesia was 2.83 million hectares in 1997-2000 (Badan Planologi Kehutanan 2008). The estimation on deforestation rate in Indonesia in 2000 – 2005 showed the value of 1.08 million hectares forest loss annually. As a result, Papua region has suffered from deforestation on 4.15% of national forest cover. This area may equal to 1.81 % of Papuan forest or about 31.773,063 ha (BPKH 2010). It was also argued that Papua basically has lost more than 100,000 ha of its forest area due to deforestation and forest degradation (Kapisa 2004 *in* Anggraeni 2007). Unfortunately, forest degradation and deforestation in Papua take place in all forest categories including conservation forest and protection forest.

The presence of Special Autonomy Regulation for Papua (Acts of Republic Indonesia, 2001) can support the growth and acceleration on local development to reach equal improvement like in other places in Indonesia. But, this regulation gives more opportunities to uncontrolled exploitation on natural resources. All of the regulation and developments may influence the communities in using their forests and increase biodiversity devastation in Papua (Frazier 2007). The threats on avifauna in Papua mainly consist of (a) bird harvesting in traditional and subsistent way, (b) timber logging for industry, (c) conversion of farmlands, (d) bird commercial trade and (e) introduction of exotic species. As the result, the existence of many endemic Papuan avifauna including *Goura victoria* in their habitats are endangered.

Based on IUCN guidance (IUCN 2001), conservation status of a species can be validated based on some categories and criterion (IUCN 2001). Those criterion included, population size, sub population, adult stage, generation, reduction, continuing decline, extreme fluctuation, severely fragmented, extend of occurrence, area of occupancy, location and quantitative analysis of the species. According to definitions of the criterion, a species can be classified into nine categories, where *Goura victoria* is classified as Vulnerable (Vu A2cd+3cd+4cd, appendix 8) and this category has not been changed since 1994 (IUCN 2012). The category was given to the bird, because its population is suspected to rapidly decline due to hunting, and degradation or/and loss of its lowland forest habitat. Recently, there had been a proposal to upgrade the status of genus *Goura* from Appendix II to Appendix I CITES, but this proposal was withdrawn in 1992 due to the lack of information on the population of *G.victoria* (King and Nijboer 1994).

The total area of tropical lowland forest in Papua is 176,750 km^2 (Johns *et al* 2007^a and Pratito Puradyatmika, *pers.comm*). This area equals to about 72.03% of total forested land in Papua Province. The area of lowland tropical rain forest in the northern part of Papua is 83.595,79 km² or about 47.30% from total area of lowland rainforest in Papua (BPKH 2010). In other words, about 44.70% (78,999.20 km²) of the total area in Papua is northern forest area. Around 2.60% (4,596.59 km²) of total Papuan forest located in Supiori, Biak and Yapen islands (appendix 6).

The habitat of *G.victoria* is lowland forests and the swampy areas between 0 - 600 m above sea level in the northern part of Papua and its satellite islands (Beehler *et al* 1986, IUCN 2012). This study has estimated that the population size of *Goura* is around 2451– 5178 birds in Buare, Unurumguay and Bonggo, and about 3344 to 6206 birds in Supiori.

These population sizes might be the latest data for Papua Indonesia, because there is no information from any other study about population of *G.victoria* in the lowland forest of northern Papua since 20 years (King and Nijboer 1994). However, these values merely originated from four small areas, and can not be generalized for the whole northern lowlands in Papua. However, the values are needed as comparative basis for other researches in the future. It is considered that the revision for conservation status of *G.victoria* is not yet needed, therefore the use of the IUCN status of "Vulnerable A2cd+3cd+4cd" and CITES status "Appendix II" can still be maintained.

5.7 Recommendations for conservation of *Goura victoria*

There are some recommendations in order to conserve *G.victoria* with sustainable persistence, as can be listed below.

1. The efforts on collection and documentation of research on *G.victoria* should be carried out in the other part of Papua, both short-terms and long-terms. The research should include all ecological aspects of *G.victoria*, such as breeding biology, reproduction, habitat preference of *Goura*, and population aspect using different methods of study. These efforts can reduce a dependence on data and information from

private-foreign research institutions, specifically from Papua New Guinea (Supriatna 1997 and 2008 also Mack and Dumbacher 2007).

2. In order to increase skills, knowledge and ability of researchers in Papua, specifically at university level, more trainings and capacity building are necessary. Recently, Papua still has deficiency in experts and scientists for Papuan birds (Beehler 1995 *in* Mack and Dumbacher 2007).

3. There is a need to develop and to intensify utilization of bird research facilities in Papua. The facilities such as Bird Park on Biak and some Animal Sanctuaries around Papua can be used as research sites for several aspects like breeding season, food preference, bird behavior, captivity efforts and other specific research related to all species of *Goura*. The research can involve undergraduate students from the universities in Papua. The facilities can be improved to build up more comprehensive data and used for species ecotourism sites.

4. Governments and related stakeholders should enhance and determine the conservation areas with the factual boundaries, including protected forest, animal sanctuaries and nature reserves. These efforts are not only particularly required in the new-developed regencies such as the Regency of Sarmi, Supiori and Mamberamo Raya, but also in other regencies that already have conservation area. These efforts can be directed to protect the presence of primary forests as habitat for animal and plants, and also to prevent the destruction and irresponsible utilization of natural resources.

5. It is essential to establish and manage more buffer zones around protected area immediately, to reduce interference from local people. It is expected that people can practice hunting and gathering non timber forest products in certain buffer zone areas without disturbing the main habitat of *Goura*. The zone systems with defined buffer zones can be considered as wise steps in conservation efforts (Makabori 2005, Supriatna 2008, O'Brien *et al* 2010).

6.   Papuan people need more sosialization of the laws and regulations concerning wildlife protection. The law enforcements should be implemented together with strict sanctions. This condition can reduce the law violation, specifically that related to the hunting practices on protected animals.

7.   Traditional and local knowledge in utilization of *G.victoria* or *Goura* spp should be further studied and assessed. This effort should include studying the local languages and cultures, and also involving local people and other related stakeholders in establishing alternative working programs for conserving natural resources and protecting wildlife species in Papua. The entire works can become important and positively valued due to the valuable property of forest and land as mother and source of life for local people in Papua.

8.   There is an urgent need to increase and to strengthen the networking with many scientific institutes, for instances CIFOR, WWF-Papua Region, International and National NGOs, LIPI and also with local people or villagers from different areas in Papua. These collaborations are supposed to build up and enlarge more efforts on participatory mapping of traditional people and their resource utilization, at their own area (Boissiére *et al* 2004, Supriatna 2008, Padmanaba *et al* 2012).

5.8 Application of the research methods

5.8.1 Assessing wildlife population methods

The estimation of population density of *G.victoria* presented here is the result from field research conducted by the researcher in certain location using distance sampling method. In this research the number of birds detected did not reach the required number for data analyzing using distance software (see Buckland *et al* 2001).   Although the amount of detections would not recommend to applying the distance sampling analysis for each study site, the Distance Sampling Program enabled the analysis of data combined from all locations using MCDS (Multi Covariate Distance Sampling) analysis. The MCDS analysis might produce the rough estimation values of the population density of *G.victoria* in each study site, and this value was furthermore used to estimate the population size of *G.victoria*.

For the future directions, improved research effort on the population density of *G.victoria* is needed. Several points should be taken into consideration specifically in setting aside the wider study area, more number of transect lines, replication of transect lines and the transect efforts.

5.8.2 Assessing hunting sustainability

This research is the first attempt to assess whether hunting activities of *Goura victoria,* carried out by Papuan people in the northern part of Papua may affect the population of *G.victoria*. For this purpose, data of the age of first reproduction and the age of last reproduction of *G.victoria* are important to estimate the maximum annual production and maximum sustainable yield (Bennett and Robinson 2000[b], Robinson 2000, Peres 2000, Begazo and Bodmer 1998). These data are not generated directly from this study and also not obtained from the study on *G.victoria* population in its natural habitat, but were collected from long-term research in captivity (more than 30 years) in Rotterdam Zoo, Netherland (King and Damen 2004, Beltermann and Poot 2008). Data from other studies and from the literatures (Coates 1985, Baptista *et al* 1997 Gibbs *et al* 2001, Pangau and Noske 2010) were also included. The reproduction data of wild animal could be used from captive populations or other literature, if the reproduction data of studied species is unavailable from their natural habitat (Robinson and Redford 1986, Noss 2000, Lee 2000, Robinson 2000).

For this research, the application and use of captivity data of *G.victoria* (Beltermann and Poot 2008), and other reproduction and hunting data (Coates 1985, Baptista *et al* 1997; Bennett and Robinson 2000[b], Robinson 2000, Gibbs *et al* 2001, Pangau and Noske 2010) have been combined with the result of this research. The results showed that traditional hunting in several location of Papua were unsustainable and have a negative effect on the population of *G.victoria*. Basic studies such as breeding biology and reproduction of *G.victoria* need to be done by the Papuan researcher to reduce the dependence on basic data from other institutions (Beehler *et al* 1995 *in* Mack and Dumbacher 2007). Besides that, in order to produce better information and recommendations for protecting a wildlife animal, it is important to have detailed data from economic, nutritional and ecological studies (Lee 2000). It is also crucial to understand the dynamics between rural economies, health and nutrition conditions of the local communities, including the biology and status of wildlife population related to hunting activities (Lee 2000).

5.8.3 Questionnaire survey

Survey research using questionnaires and interviews (e.g. semi-structural interview) are commonly carried out by social and anthropology scientists (O'Brien *et al* 1998). They usually apply these methods in the study of ethnobotany (Ayatunde *et al* 2008), ethnozoology (Alves and Rosa 2010), and participatory mapping (Boissiere *et al* 2004, Boissiere *et al* 2007, Padmanaba *et al* 2012). Basically, these methods also can be used for digging up information on hunting from different aspects, for instance to collect data through the local knowledge about the existence of wildlife animal in a region, to investigate the motivation and preference of hunting activities, or to explore and describe bushmeat market (Hard and Upoki 1997, O'Brien *et al* 1998, Pailler *et al* 2009). Eventhough, these methods are rarely used in researches about tropical wildlife (O'Brien *et al* 1998), few studies (Pangau-Adam *et al* 2012) including this study have already used the questionnaire surveys in Papua.

The main problem of this survey was the different understanding of each respondent to the questions. Usually, the researcher should ask again the question or give more explanation to the respondents in order to get their answer. In some cases, the answer were delivered in "unclear" meaning, for example when the respondent had to answer the concept of rare, commonly or occasionally, or between easy versus difficult. It happened, because according to the respondent these terms had similar meaning. To minimize the problem, questions were asked simply and implicitly in more relaxed situation, so the respondent can give the answer without any distrustful impression. In particular, close approaches to local people in the study sites were needed, because the basic character of Papuan people is commonly timid, reclusive and shy when they met 'foreign people' even the people from other part of Papua (*pers. obsv*). Another important issue to be noted is how the researcher should determine the study site, respondents and the appropriate field helper appropriately. Therefore, the researcher should have sufficient advices and assistances from the chief of districts, the leader of villages and *Ondoafi* in making decision regarding research site and respondent. The information on hunting areas, the tribes and the numbers of hunters had also to be taken into account.

Apart of many obstacles mentioned above, it is documented that research based on survey and interview with local hunters is more effective when applying rapid survey methods on large regions (Hard and Upoki 1997, O'Brien *et al* 1998). This method can gather more information from local people.

Generally, this survey on *Goura* hunting can simply describe the hunting pattern by local hunters in four different areas only, and the result and conclusion of this study is not applicable to all hunters in northern Papua. However, the information from this research can be considered as the basic and preliminary data for the future research in the field of biodiversity management and conservation of endemic species in Papua.

Chapter 6: CONCLUSION

Four important topics have been studied through this research, and those are (1) estimation density of *Goura* population, (2) forest structure of *Goura* habitat, (3) hunting activity of Papuan people and its impacts on the *Goura* population and (4) conservation efforts on *G.victoria*.

This study was carried out in four sites in different regencies of Papua. Three sites (Unurumguay-Jayapura, Bonggo-Sarmi and Buare-Mamberamo Raya) were located in the northern of Papua, and Supiori site in Supiory Island Regency was located in the gulf of Cenderawasih. Estimation density of *Goura* population in each study site was done through bird surveys using transect lines and distance sampling. Using questionnaires, a total of 151 respondents from 13 villages were interviewed to reveal the hunting activities on *G.victoria*. Information on habitat structure and vegetation composition was collected in each site using randomly line transect for vegetation analysis.

The results may be concluded as follow:

The composition and tree diversity of forest where *Goura* occurred varies among all study sites. A total of 58 tree species in 38 families were encountered in Buare forest (Mamberamo Raya Regency); 57 species in 38 families in Supiori forest (Supiori Regency), 39 species in 25 families in Unurumguay forest (Jayapura Regency) and 34 species in 22 families in Bonggo forest (Sarmi Regency).

Vegetation in all study sites were dominated by trees with diameter at breast height or *dbh* of 10 - 34 cm. Trees height in all study sites were ranging from 6 m to 25 m, dominated by trees height between 6 m to 15 m. The forest in all sites can be classified as forest with mid-lower canopy.

Comparing the tree composition, it was found that Shannon-Wiener Diversity Index (H') was 3.55 in Buare, 3.45 in Supiori, 3.09 in Unurumguay and 3.06 in Bonggo. All indexes were in range of 1 - 4.5, means all study sites was classified as forests areas with higher tree diversity and abundance.

The result of ANOVA on Shannon-Wiener Diversity Indexes between each sites (in total 6 combinations) with t-test (95% confidence interval) showed that the forest in Buare, Supiori and Unurumguay had higher tree diversity than Bonggo forest.

Estimation on population density of *Goura victoria* was different among the study sites. A total of 41.8 individuals per km^2 were estimated inhabiting Buare forest, 40.3 individuals per km^2 in the Supiori, 30.8 individuals per km^2 in Unurumguay forest and about 13.1 individuals per km^2 in Bonggo forest. It is found that the estimation density of *Goura* populations different among study sites, and Bonggo forest has the lowest population of *Goura victoria*.

The estimated population size of *G.victoria* per hunting area was varied between each study site. It was about 2,451 – 4,380 birds in Buare hunting area, 3,344 – 6,206 birds in Supiori and 2,589 – 4,682 birds in Unurumguay while 3,266 – 5,178 birds in Bonggo hunting area.

Papuan hunters commonly apply non-selective system in their hunting practice and they mainly use trapping techniques and some modifications in wildlife hunting. Recently, the hunting practices are not only subsistent, but also for commercial purposes. The statistical analysis showed a significant correlation between hunting using air gun and the number of captured *Goura* in Bonggo and Supiori sites.

Hunting on *G.victoria* in each hunting area was unsustainable, because the estimated current harvest levels exceed the estimated maximum sustainable annual harvest. It means hunting activities of Papuan hunters have negative impacts on the population of *G.victoria* population, although they mostly used the simple hunting equipments like foot snares.

Concerning the management of wildlife hunting in Papua, local government should focus on the socialization of hunting laws and implementation of law enforcement, as well as controlling illegal hunting and wildlife trade. In several regions of Papua, local people have such customary regulation, traditional knowledge and wisdom such as taboos and sacred places. Therefore, biodiversity management and forest conservation in Papua should include comprehensive involvements of local communities, traditional management and customary rights, and support from government.

REFERENCES

Agerloo, M. and W.R.J. Dekker. 1996. Exploitation of megapode eggs in Indonesia: the role of traditional methods in the conservation of megapode. *Oryx* **30** (1):59-64

Aiyadurai, A., N.J. Singh and E.J. Milner-Gulland. 2010. Wildlife hunting by indigenous tribes: a case study from Arunachal Pradesh, North-east India. *Oryx* **44** (4): 564-572

Aiyadurai, A. 2011. Wildlife hunting and conservation in northeast India: a need for an interdisciplinary understanding. *International Journal of Galliformes Conservation* **2**: 61-73. http://www.pheasant.org.uk/uploads/Pages61-73_Aiyadurai.pdf (accessed January 16, 2013)

Alamgir, M., Md.S.H. Sarkas, Md.S.I. Sohel and S. Akhter. 2011. Tree diversity and biodiversity conservation potentials in Khadimnagar National Park of Bangladesh. *Tigerpaper* **38** (2): 20-28 http://www.fao.org/docrep/014/am635e/am635e00.pdf (accessed April 9, 2012)

Alvard, M. 2000. The impact of traditional subsistence hunting and trapping on prey population: data from Wana horticulturists of Upland Central Sulawesi,Indonesia. **In:** *Hunting for sustainable in tropical forest*, ed. J.G. Robinson and E.L. Bennett. 214-232. New York. Columbia University Press.

Alves, R.R.N. and I.L. Rosa. 2010. Trade of animal used in Brazillian traditional medicine: Trends and implications for conservation. *Human Ecology* **38**: 691- 704. DOI 10.1007/s10745-010-935-0

Andrew. 1992. *The Bird's of Indonesia. A Check List (Peter's Sequence).* Kukila Check List No. 1. Indonesian Ornithological Society.

Andoy, E.E.S. 2002. Study populasi rusa (*Cervus timorensis*) dan perburuan oleh penduduk di desa Poo, Tomer dan Sota dalam Taman Nasional Wasur, Merauke. (Skripsi sarjana) Manokwari, Papua Barat: Universitas Negeri Papua.

Anggraeni, D. 2007. Pattern commercial and industrial resource use in Papua. **In** *The Ecology of Papua Part Two*, Ed A.J. Marshall and B. Beehler. 1149-1166. Singapore. Periplus Edition.

Anggraini, K., M. Kinnaird, T. O'Brien. 2000. The effects of fruit availability and habitat disturbance on an assemblage of *Sumatran hornbills*. *Bird Conservation International* **10**:189-202

Apomfires, F. 2002. Makanan pada komunity adat Jae: catatan sepintas lalu dalam penelitian gizi. *Anthropologi Papua* **1** (**2**): 1-9

Apomfires, F. and K. Sapulette. 1992. Masyarakat Arfak di Anggi, Kabupaten Manokwari. **In**: *Irian Jaya Membangun Masyarakat Majemuk.* Seri Etnografi Indonesia 5, ed Koentjaraningrat. 139-155. Jakarta. Penerbit Djambatan.

Ariantiningsih, F. 2000. Sistem perburuan dan sikap masyarakat terhadap usaha-usaha konservasi rusa (*Cervus timorensis*) di pulau Rumberpon Kecamatan Ransiki Kabupaten Manokwari (Skripsi sarjana). Manokwari, Papua Barat: Universitas Negeri Papua.

Ayatunde, A.A., M. Briejer, P. Hiernaux, H.M.J. Udo and R. Tabo. 2008. Botanical knowledge and its differentiation by age, gender and ethnicity in south Niger. *Human Ecology* **36**: 881-889. DOI 10.1007/s10745-008-9200-7

Badan Perencanaan dan Pembangunan Nasional (BAPPENAS). 2003. Indonesian Biodiversity and Action Plan for 2003-2020. Ministry of National Development Planning. National Development Planning Agency. Jakarta.

Badan Planologi Kehutanan. 2008. Rekalkulasi Penutupan Lahan Indonesia Tahun 2008. Pusat Inventarisasi dan Perpetaan Hutan. Departemen Kehutanan Republik Indonesia. Jakarta. http://forestclimatecenter.org/files/2008%20Rekalkulasi%20Penutupan%20Lahan%20Indonesia%20Tahun%202008.pdf (accessed August 5, 2013)

Badan Pusat Statistik (BPS). 2010[a]. Hasil Sensus Penduduk Propinsi Papua. 2010 . Data agregat per Kota dan Kabupaten. Badan Pusat Statistik Indonesia. - Provinsi Papua. Jayapura. http://sp2010.bps.go.id/files/ebook/9400.pdf (accessed April 11, 2012)

Badan Pusat Statistik (BPS). 2010[b]. Papua Dalam Angka 2010 (*Papua in figure 2010*). Number of catalog: 1102001.94. Badan Pusat Statistik Indonesia - Provinsi Papua. Jayapura http://papua.bps.go.id/site/index.php?option=com_docman&task=cat_view&gid=196&Itemid=27 (accessed April 11, 2012)

Badan Pusat Statistik (BPS) Jayapura. 2010. Jayapura dalam angka 2010 (Jayapura Regency in figures 2010 Number of catalog: 1102001.94. Badan Pusat Statistic Indonesia-Kabupaten Jayapura. http://jayapurakab.bps.go.id (accessed July 1, 2012)

Badan Pusat Statistik (BPS) Biak-Numfor. 2012. Biak-Numfor dalam angka 2011/2012 (Biak-Numfor Regency in figures 2011/2012). Number of catalog: 9409.1201. Badan Pusat Statistic Indonesia-Kabupaten Biak-Numfor. http://de.scrib.com/doc/139271816/Biak-dalam-angka-2012) (accessed August 5, 2013)

Balai Pemantapan Kawasan Hutan Wilayah X (BPKH) Jayapura. 2008. *Statistik Kehutanan Propinsi Papua Tahun 2008*. Direktorat Jenderal Planologi Kehutanan. Kementrian Kehutanan Indonesia. Papua.

Balai Pemantapan Kawasan Hutan Wilayah X (BPKH). 2010. Suplemen data luas penutupan lahan pada fungsi kawasan hutan propinsi Papua (*Data supplement of lands cover areas based on forest function in Papua Province*). Direktorat Jenderal Planologi Kehutanan. Kementrian Kehutanan Indonesia. Papua.

Baptista, L.F., P.W. Trail and H.M. Horblit. 1997. Order Columbiformes, Family Columbidae (Pigeons and Doves) **In**: *Handbook of the Birds of the World. Volume 4 Sangrouse to Cuckkos*. ed. J. del Hoyo, A. Elliot and J. Sargatal . Barcelona. Lynx Edicions.

Barrio, J. 2011. Hunting Pressure on Cracids (Cracidae: Aves) in forest concession in Peru. *Rev.peru.biol.* **18 (2)**: 225-230

Bird, R.B. and D.W. Bird. 2008. Why women hunt, Risk and contemporary foraging in a Western desert Aboriginal community. *Current Anthropology* **49 (4)**: 655-692

BirdLife International 2012. *Threatened Birds of Asia: the Bird Life International Red Data Book*. BirdLife International, Cambridge. http://www.birdlife.org/datazone/info/spcasrdb (accessed 31 October 2012).

BirdLife International. 2012. Endemic Bird area of Northern Papua lowlands http://www.birdlife.org/datazone/ebafactsheet.php?id=176 (accessed July 12, 2012).

BirdLife International. 2012. Endemic Bird area of Northern Papua lowlands http://www.birdlife.org/datazone/ebafactsheet.php?id=174 (accessed July 12, 2012).

Beehler, B. 1982. Ecological structuring of forest bird communities in New Guinea. **In**: *Monographiae Biologicae*, Volume **42.** ed. J.L. Illies, J. and F.R.G. Schlitz in *Biogeography and Ecology of New Guinea* Volume **1.** ed. J.L. Gressitt. 837- 861. London. Dr. W. Junk Publishers.

Beehler, B., T.K. Pratt and D.A. Zimmerman. 1986. *The Birds of New Guinea.* New York Princeton University Press. .

Begazo, A.J. and R.E. Bodmer. 1998. Use and conservation of cracidae (Aves: Galliformes) in the Peruvian Amazon. *Oryx* **32** (**4**): 301-309

Bennett, E.L. and J.G. Robinson. 2000[a]. Hunting for the snark. **In:** *Hunting for sustainable in tropical forest,* ed. J.G. Robinson and E.L. Bennett. 1-9. New York. Columbia University Press.

Bennett, E.L. and J.G. Robinson. 2000[b]. Hunting for sustainability: The start of synthesis. **In:** *Hunting for sustainable in tropical forest*, ed. J.G. Robinson, and E.L. Bennett. 499-520. New York. Columbia University Press.

Bird, R.B. and D.W. Bird. 2008. Why women hunt. Risk and contemporary foraging in a Western desert Aboriginal community. *Current Anthropology* **49** (**4**): 655-692.

Bodmer, R.E. 1995. Managing Amazonian Wildlife: Biological Correlates of Game Choice by Detribalized Hunters. *Ecological Applications* **5** (**4**): 872-877

Bodmer . R.E., J.F. Eisenberg and K.H. Redford. 1997. Hunting and likelihood of extinction of Amazonian mammals. *Conservation Biology* **11** (**2**): 460-466

Boissiére, M., N. Liswanti, M. Padmanaba and D. Sheil. 2007. *People priorities and perception-towards conservation partnership in Mamberamo.* Center for International Forests Research. Bogor. Indonesia. http://www.cifor.org/mla/download/publication/People%20priorities.pdf (accessed April 9, 2012)

Boissiére, M., M. van Heist, D. Sheil, I. Basuki, S. Frazier, U. Ginting, M. Wan, B. Hariadi, H. Hariyadi, H.D. Kristianto, J. Bemey, R. Haruway, E.R.Ch. Marien, D.P.H. Koibur, Y. Watopa, I. Rachman and N. Liswanti. 2004. Pentingnya Sumber daya alam bagi masyarakat lokal di daerah aliran sungai Mamberamo, Papua dan implikasinya bagi konservasi. *Journal of Tropical Ethnobiology* **1** (**2**) 76-95 http://www.cifor.org/publications/pdf_files/articles/ABoissiere0401.pdf (accessed April 9, 2012)

Brickle, N.W., J.W. Duckworth, A.W. Tordoff, C.M. Poole, R. Timmins and P.J.K. McGowan. 2008. The status and conservation of Galliformes in Cambodia, Laos and Vietnam. *Biodiversity and Conservation* **17**: 393-1427 DOI 10.1001/S10531-008-9346z

Brower, J.E., J.H. Zarr and C. von Ende. 1997. *Field and Laboratory Methods for General Ecology* 2nd Edition. USA. Brown Publisher.

Buckland, S.T., D.R. Anderson, K.P. Burnham, J.L. Laake, D.L. Borchers and L. Thomas. 2001. *Introduction to distance Sampling. Estimating abundance of biological population.* New York. Oxford University Press.

Buiney, E.T. 2006. Perburuan dan perdagangan jenis-jenis burung disekitar kawasan pemukiman dan hutan distrik Nimbokrang Kabupaten Jayapura (Skripsi sarjana). Jayapura, Papua: Universitas Cenderawasih. .

Bulmer, R. 1968. The strategies of hunting in New Guinea. *Oceania* **38** (**4**): 302-318.

Cahya, D. N. 2000. Teknologi berburu rusa (*Cervus timorensis*) dan kasuari (*Casuarius* sp) secara tradisionil pada suku Marind dan Kanuum di kawasan Taman Nasional Wasur Kabupaten (Skripsi sarjana). Manokwari, Papua Barat: Universitas Cenderawasih.

Castelleta, M., N.S. Sodhi and R. Subaraj. 2000. Heavy extinctions of forest avifauna in Singapore: lessons for biodiversity conservation in Southeast Asia. *Conservation Biology* **14**: 1870-1880

Castelletta, M., J.M. Thiolay and N.S. Sodhi. 2005. The effects of extreme forest fragmentation on the bird community of Singapore Island. *Biological Conservation* **121**: 135-155

Centre for Research in Ecological and Environmental Modeling (CREEM). 2009. International Workshops: Introduction to Distance Sampling. St. Andrews 18th-21th August 2009. England. University of St. Andrews.

Chamberlain, J.L., A.B. Cunningham and R. Nasi. 2004. Diversity in forest management: non forest timber products and bushmeat. *Renewable Resources Journal* **22**(**2**): 11-19

Cinner, J.E. 2007. The role of taboos in conserving coastal resources in Madagascar. *SPC Traditional marine Resource management and Knowledge Information Bulletin#***22**: 15-23.

CITES. 2012. Amendments to appendices I and II of the Convention: others proposals http://www.cites.org/eng/cop/08/prop/E08-Prop-24_Goura.PDF (accessed October 12, 2012

Clayton, L. and E.J. Milner-Gulland. 2000. The trade in wildlife in North Sulawesi, Indonesia. **In** *Hunting for sustainable in tropical forest*, ed. J.G. Robinson and E.L. Bennett. 473-498. New York. Columbia University Press.

Cleary, D.F.R., T.J.B. Boyle, T. Setyawati, C.D. Anggraei, E.E. van Loon and S.B.J. Menken. 2007. Bird species and traits associated with logged and unlogged forest in Borneo. *Ecological Application* **17** (**4**): 1184-1197

Coates, B.J. 1985. *The Birds of Papua New Guinea, Including the Bismarck Archipelago and Bouganville.* Volume 1. Non-Passerine. Alderly-Queensland. Dove Publications.

Collar, N.J., M.J. Crosby and A.J. Statterfield. 1994. *Bird to Watch 2, The World List of Threatened Birds-The Official Source for Birds on the IUCN Red List.* Bird Life Conservation Series No. 4. BirdLife International.

Colell, M., C. Maté and J. E. Fa. 1994. Hunting among Moka Bubis in Bioko: dynamics of faunal exploitation at the village level. *Biodiversity and Conservation* 3: 939-950

Corlett, R.T. 2009. *The Ecology of Tropical East Asia.* Oxford University Press. New York. 262 Pages

Corlett, R.T. 2007. What's so special about Asian tropical forests. *Current Science* 93 (11): 1551-1557

Cuthbert, R. 2010. Sustainability of hunting population densities, intrinsic rates of increase and conservation of Papua New Guinean mammals: a quantitative review. *Biological Conservation* 143: 1850-1859.

Dumatubun, A.E. and K. Wanane. 1992. Masyarakat Senggi , di dekat batas timur negara. **In**: *Irian Jaya Membangun Masyarakat Majemuk.* Seri Etnografi Indonesia 5, ed. Koentjaraningrat. 297-312. Jakarta. Penerbit Djambatan.

Dumbois, D.M. and H. Ellenberg. 1974. *Aims and methods of vegetation ecology.* New York. John Willey and Sons.

Dwyer, P.D. and M. Minnegal. 1991. Hunting in lowland, tropical rain forest: towards a model of non agricultural subsistence. *Human Ecology* 19 (2): 187-212.

Dwyer, P.D. 1974. The price of protein: five hundred hours of hunting in the New Guinea Highlands. *Oceania* 44 (4): 278-293.

Eden, M.J. 1993. Swedden cultivation in forests and savanna in lowland Southwest Papua New Guinea. *Human Ecology* 21 (2): 145-166

Ellis, E.C., E.C. Antil, H. Kreft. 2012. All is not loss: Plant biodiversity in the Anthropocene. *PLoSONE*: e30535.doi 10.1371/journal.pone.0030535 http://www.plosone.org/article/info%3Adoi%2F10.1371%2Fjournal.pone.0030535 (accessed January 23, 2012)

Elsham Papua. 2008. Bupati Mamberamo Raya: HPH tidak beri kontribusi. http://elshamnewsservice.wordpress.com/2008/04/16/bupati-mamberamo-raya-perusahaan-hph-tak-beri-kontribusi/ (accessesd 5 September, 2013).

Escamillia, A., M. Sanvicente, M. Sosa., and C. Galinido-Leal. 2000. Habitat mosaic, wildlife ability, and hunting in tropical forest of Calakmul, Mexico. *Conservation Biology* **14** (**6**): 1592-1601

Fa, J.E. 2000. Hunted animals in Bioko Island, West Africa: Sustainability and future. **In:** *Hunting for sustainable in tropical forest*, ed. J.G. Robinson and E.L. Bennett. 1-9. New York. Columbia University Press.

Fa, J.E. and D. Brown. 2009. Impact of hunting on mammals in Africal moist forest: a review and synthesis. *Mammal Rev.* **39** (**4**): 231-264

Fachrul, M.F. 2008. Metode sampling bioekologi. Jakarta. PT. Bumi Aksara.

Farida, W.R., G. Semiadi, Widarteti and H. Dahruddin. 2001. Pemanfaatan kuskus (Phalanger sp.) oleh masyarakat Timor Barat, Nusa Tenggara Timur (*Utilization of cuccus (Phalanger sp.) by communities of Timor Barat, Nusa Tenggara Timur*). *Biota* **6** (**2**): 85-86

Fatem, S., J. van Laar, J.L. Sroyer and J. Manusaway. 2011. Portrait of community mapping stage through zoning system on management of Teluk Cenderawasih National Park, West Papua. *Tigerpaper* **38** (**2**): 13-19 http://www.fao.org/docrep/014/am635e/am635e00.pdf (accessed January 14, 2013).

Fowler, Jim and L. Cohen. 1995. *Statistics for Ornithologists* 2nd Edition. BTO Guide 22. British. British Trust for Ornithology.

Forest Watch Indonesia/Global Forest Watch. 2001. Keadaan hutan Indonesia (*Indonesia forests situation*). Forest Watch Indonesia and Global Forest Watch Washington DC. http://pdf.wri.org/indonesia_forest_matter_id.pdf (accessed January 18, 2013)

Franzen, M. 2006. Evaluating sustainability of hunting comparison of harvest profiles across three Huorani communities. *Environmental Conservation* **33** (**1**): 36-45

Frazier, S. 2007. Threats to biodiversity. **In:** *The Ecology of Papua Part Two*, eds. A.J. Marshall, and B. Beehler. 1199-1129. Singapore. Periplus Edition.

Frith, H.J., F.H.J. Crome and T.O Wolfe. 1976. Food of fruit pigeons in NewGuinea. *The Emu* **76** (**2**): 49-58

Garcia, G. and S.M. Goodman. 2003. Hunting protected animals in the Park National d'Ankara Fantsika, North western Madagaskar. *Oryx* **37**: 115-118

Geneleti, D. 2003. Biodiversity impact assessment of road: an approach based on ecosystem rarity. *Environmental Impact Assessment Review* **23**:343-365. DOI10.1016/S0195- 9255(02)00099-9

Gibbs, D., E. Barnes and J. Cox. 2001. *Pigeons and doves a guide to the pigeons and doves of the World.* London. Christopher Helm.

Gillison, A.N., N. Lisnawati, S. Budidarsono, M. van Noordwijk and T.P. Tomich. 2004. Impact of cropping methods on biodiversity in coffee agroecosystem in Sumatra, Indonesia. Ecology and Society 9 (2): 1-31 online URL: http://www.ecoloyandsociety.org/vol11/iss2/art20 (accessed October 31, 2012).

Golden, C.D. 2009. Bushmeat hunting and use in Makira Forest, north-eastern Madagascar: a conservation and livelihood issue. *Oryx* **43(3)**: 386-392. DOI 10.1017/50030605309000131

Goodwin, D. 1977. *Pigeon and Doves of The World 2nd Edition.* Ithaca. Cornell University Press.

Goodman, M.J. and P.B. Griffin. 1985. The compatibility of hunting and mothering among the Agta hunter-gatherers of the Philippines. *Sex Roles* **12 (11/12)**: 1199 –1209

Griffin, P.B. and M.B. Griffin. 2000. Agta hunting and sustainability of resource use in Northeastern Luzon, Philippines. **In:** *Hunting for sustainable in tropical forest,* eds. J.G. Robinson and E.L. Bennett. 325-335. New York. Columbia University Press.

Hadi, S., T. Siegler, M. waltert and K. Jones. 2009. Tree diversity and forest structure in northern Siberut, mentawai island Indonesia. *Tropical Ecology* **50 (2)**: 315-327

Handini, S., S.P. Waluyo, dan J. Manansang. 1992. Palatabilitas burung *Goura victoria* (F) terhadap bermacam-macam pakan di kandang penangkaran Taman Safari Indonesia. *(Palatability of crowned pigeon Goura victoria (F) on various food items in captivity, at Indonesian Safari Park Bogor).* Prosiding Seminar Hasil Penelitian Litbang SDH. 57-62. Bogor. Puslitbang Biologi LIPI.

Hard, J.A. and A. Upoki. 1997. Distribution and conservation status of Congo Peafowl *Afropavo congensis* in eastern Zaire. *Bird Conservation International* **7 (4)**: 295-316.

Hardjanti, F. 2005. Convention on International Trade in Endangered Species/CITES. Makalah dalam Prosidings Judicial Workshop Penegakan Hukum Atas Perlindungan Satwa Liar. Tobing, H., F. Hanif dan R. Setyaningrum (Editors). Kerjasama WWF Indonesia-TRAFFIC South Asia- Mahkamah Agung Republik Indonesia. Cibodas 27-29 September 2005. http://rafflesia.wwf.or.id/library/admin/attachment/books/prosidings%20 versi%20pdf.pdf   (accessed January 17, 2013).

Haugaasen, T. and C.A. Peres. 2008. Vertebrate responses to fruit production in Amazonian flooded and unflooded area. *Biodiversity Conservation* **16**: 4165-4190. DOI. 10.1007/s10531-007-9217-z

Healey, C.J. 1978. Effects of human activity on *Paradisea minor* in the Jimmy Valley, New Guinea. *Emu* **78**: 149-155.

Hope, G.S. 2007. The history of human impact on New Guinea. **In**: *The Ecology of Papua Part Two*, ed. A.J. Marshall and B. Beehler. 1087-1097. Singapore. Periplus Edition.

Hyndman, D.C. 1984. Hunting and classification of game animals among the Wopkaimin. Oceania **54 (4)**: 289-309

International Union for Conservation of Nature and Natural Resources. 2001. IUCN Red List Categories and Criteria: Version 3.1. IUCN Species Survival Commission, IUCN, Gland, Switzerland and Cambridge, UK. Ii +30 pp. Available from: http://www.iucn.org/themes/ssc/redlists/RLcast2001booklet.html. (accessed October 8, 2012).

International Union for Conservation of Nature and Natural Resources. 2009. The IUCN Red List of threatened species: *Paradisea minor* http://www.iucnredlist.org/details/106005842/0   (accessed January 30, 2013).

International Union for Conservation of Nature and Natural Resources. 2012. The IUCN Red List of threatened species: *Goura victoria* http://www.iucnredlist.org/details/106002755/0 (accessed November 21, 2012).

Iridale, T. 1956. *Birds of New Guinea*. Melbourne. An Australiana Society Publication.

Jepson, P., R.J. Ladle and Sujatnika. 2011. Assessing market-based conservation governance approaches: a socio-economic profile of Indonesian market for wild bird. *Oryx* **45 (4)**: 482-491. DOI 10.1017/S003060531100038x

Johns, A.D. 1983. Tropical forest primates and logging-can they co-exist? *Oryx* **17 (3)**: 114 – 118

John, A.D. 1985. Selective logging and wildlife conservation in tropical rain-forest: Problem and recommendations. *Biological Conservation* **31**: 355-375

Johns, A.D. and B.G. Johns. 1995. Tropical forest primates and logging: long-term coexistence? *Oryx* **29 (3)**: 197-204

Johns, R.J., G.A. Shea and P. Puradyatmika. 2007[a]. Lowland swamp and peat vegetation of Papua. **In**: *The Ecology of Papua Part Two*, eds. A.J. Marshall and B. Beehler. 910-944. Singapore. Periplus Edition.

Johns, R.J., G.A. Shea and P. Puradyatmika. 2007[b]. Lowland vegetation of Papua. **In**: *The Ecology of Papua Part Two*, eds. A.J. Marshall and B. Beehler. 945-961. Singapore. Periplus Edition.

Johnson, A., R. Bino and A. Igag. 2004. A preliminary evaluation of the sustainability of cassowary (Aves: Casuariidae) capture and trade in Papua New Guinea. *Animal Conservation* **7**: 129-137.

Joint Nature Conservation Committee 2005. Joint Nature Conservation Committee (JNCC) Report no. 381. Checklist of birds listed in the CITES appendices and in EC regulation no. 338/97. 8th edition-UNEP-WCMC. http://jncc.defra.gov.uk/PDF/jncc381.pdf (accessed September 12, 2013)

Jorgenson, J.P. 1995. Maya subsistence hunters in Quintana Roo, Mexico. *Oryx* **29 (1)**: 49-57

Kabelen, F and M. Warpur. 2009. Struktur, komposisi jenis pohon dan nilai ekologi vegetasi di kawasan hutan kampung Sewan Distrik Sarmi, Kabupaten Sarmi. *Jurnal Biologi Papua* **1 (2)**: 72-80.

Kanninen, M., D. Murdiyarso, F. Seymor, A. Angelsen, S. Wunder and L. German. 2009. *Do trees grow on money? The implications of deforestation research for politics to promote REDD*. Forest Perspectives 4. Bogor-Indonesia. CIFOR.

Kaul, R., Hilaluddin, J.S. Jandrotia and P.J.K. McGowan 2004. Hunting of large mammals and pheasants in the Indian western Himalaya. *Oryx* **38** (**4**):426-431. DOI 10.1017/S0030605304000808

Keane, A., M.D.L. Brooke and P.J.K. McGowan. 2005. Correlates of extinction risk and hunting pressure in game birds (Galliformes). *Biological conservation* **126**: 216-233

Kilmaskossu, A. 2001. Ekologi persarangan, Musim perkembangbiakan dan kajian awal keragaman morfogenetik mambruk Polos *Goura cristata*. (master tesis), Bogor, Indonesia: PPS Institut Pertanian Bogor.

King, C.E. and J. Nijboer. 1994. Conservation Consideration for Ground Pigeons, genus *Goura*. *Oryx* **28** (**1**): 22-30

Kinnaird, M.F. and T.G. O'Brien. 1999. Breeding ecology of the Sulawesi Red-Knobbed hornbill *Acceros cassidix*. Ibis 141: 60-69

Kinnaird, M.F., T.G. O'Brien, F.R. Lambert and D. Purmiasa. 2003. Density and distribution of the endemic Seram cockatoo *Cacatua moluccensis* in relation to land use patterns. *Biological Conservation* **109**: 227-235.

Kitchener, A.C., A.A. MacDonald and P. Howard. 1993. First record of the Blue Crowned Pigeon *Goura cristata* on Seram. *Bull. Brit. Orn. Club* **113**: 43-43

Kuster, K., R. Achdiawan, B. Belcher and M.R. Perés. 2006. Balancing development and conservation? An assessment of livelihood and environmental outcomes of non forest timber product trade in Asia, Africa and America Latin. Ecology and Society 11 (2): 1-22. Online URL http://www.ecoloyandsociety.org/vol11/iss2/art20 (accessed October 31, 2012).

Kwapena, N. 1984. Traditional conservation and utilization of wildlife in Papua New Guinea. *The Environmentalist* **4**: 22-26 (*Supplement* No. 7).

Laake, Th., J.L. Laake, S. Strinberg, F.F.C. Marques, S.T. Buckland, D.L. Borchers, D.L. Anderson, K.P. Burnham, S.L. Hedley, J.L. Pollard, J.R.B. Bishop, T.A. Marques. 2006. Distance 5.0 Release 2. Research Unit for Wildlife Population Assessment, University of St. Andrews. UK. http://www.ruwpa.st.and.ac.uk/distance/. (accessed October 4, 2008).

Lambek, M. 1992. Taboo as Cultural practice Among Malagassy Speakers. *Men, New Series* **7** (**2**): 245-266.

Lamera, J. and L. Siregar. 1992. Masyarakat Bauzi di Danau Bira, Mamberamo Tengah. **In**: *Irian Jaya Membangun Masyarakat Majemuk*. Seri Etnografi Indonesia 5, ed. Koentjaraningrat. 214-229. Penerbit Djambatan.

Lee, R. J. 2000. Impact of subsistence hunting in north Sulawesi, Indonesia and conservation option. **In**: Robinson, J.G. and E.L. Benneth (Eds)

Hunting for sustainable in tropical forest. Columbia University Press. New York. Pp. 455-472.

Mack, A.L. and L.E. Alonso (Eds). 2000. A biological assessment of the Wapoga River area of Northwestern of Irian Jaya, Indonesia. Rapid Assessment Program (RAP) Bulletin Biological Assessment No. 14. Washington DC. Conservation International.

Mack, A.L. and J. Dumbacher. 2007. Birds of Papua. **In**: *The Ecology of Papua Part One*, eds. A.J. Marshall and B. Beehler. 654-688. Singapore. Periplus Edition.

Mack, A.L and P. West. 2005. Ten thousand tones of small animals: wildlife consumption in Papua New Guinea, a vital resource in need of management. Working Paper No. 61. Resource Management in Asia-Pacific.
https://crawford.anu.edu.au/rmap/pdf/Wpapers/rmap_wp61.pdf
(accessed June 13, 2013).

Madhusudan, M.D. and K.U. Karanth. 2002. Local hunting and the conservation of large mammals in India. *Ambio* **3**: 49 – 54.

Magurran, A.A. 1987. *Ecological Diversity and its Measurement*. New Jersey. Princeton Unversity Press

Mahuse, Ch. 2006. Perburuan satwa liar oleh masyarakat Genyem District Nimboran-Kemtukgresi (skripsi sarjana) Jayapura, Papua: Universitas Cenderawasih.

Makabori, Y.Y. 2005. Pergeseran pola perilaku kepatuhan masyarakat pada norma adatnya kasus pergeseran nila Igya Ser hanjop pada masyarakat lokal di kawasan cagar Alam pegunungan Arfak Kabupaten Manokwari. (Master thesis) Bogor, Indonesia: Sekolah Pasca Sarjana IPB.

Mansoben, J.R. 2005. Konservasi sumber daya alam Papua ditinjau dari aspek budaya. *Anthropology Papua* **2** (**4**): 1-12
http://www.papuaweb.org/uncen/dlib/jr/antropologi/02-04/jurnal.pdf
(accessed January 14, 2013)

Marsden, S.J. and J.D. Pilgrim. 2003. Factors influencing the abundance of parrots and Hornbill in pristine and disturbed forests on New Britain, PNG. *Ibis* **145**: 45-53

Marsden, S.J. 1998. Changes in bird abundance following selective logging on Seram, Indonesia. *Conservation Biology* **12** (**3**): 605-611

Marsden, S.J. 1992. The distribution, abundance and habitat preferences of the Salmon- crested Cockatoo *Cacatua moluccensis* on Seram, Indonesia. *Bird Conservation International* **2**: 7-14

Marthin, T.E. and G.A. Blackburn. 2012. Habitat associations of an insular Wallacean avifauna: A multi-scale approach for biodiversity proxies. *Ecological Indicator* **23**: 491-500

Mayr, E. 1941. *List of New Guinea Birds: a systematic and faunal list of the birds of New Guinea and adjacent island.* New York. The American Museum of Natural History.

McAllister, D.E., J.F. Craig, N. Davidson, S. Delamy and M. Seldon. 2001. Biodiversity impact of large dams. Background Paper No.1. Prepare for IUCN/UNEP/WCD. http://intranet.iucn.org/webfiles/doc/archieve/2001/iucn.pdf (accessed October 22, 2012)

McGowan, P.J.K. and P.J. Garson. 2002. The Galliformes are highly threatened: should we care ? *Oryx* **36** (**4**):311-312. DOI: 10.1017/S0030605302000601.

Mena V. Patricio, J.R. Stallings, J. Regaldo B., and R. Cueva L. 2000. The sustainability of current hunting practices by the Huorani. **In**: *Hunting for sustainable in tropical forest*, ed. J.G. Robinson and E.L. Bennett. 57-78. New York. Columbia University Press.

Milner-Gulland, E.J., E.L. Bennett and the SCB 2002 Annual Meeting Wild Meat Group. 2003. Wild meat: the bigger picture. *TRENDS in Ecology and Evolution* **18** (**7**): 351- 357.

Mirmanto, E. 2009. Analisa vegetasi hutan pamah di Pulau Raja Ampat, Papua (*Vegetation analysis of lowland forest in Batanta island, Raja Ampat, Papua*) *Jurnal Bioloy Indonesia* **6** (**1**): 79-96.

Miranda, M., P. Burris, J.F. Bingcang, P. Shearman, J.O. Briones, A. la Vina and S. Menard. 2003. *Mining and critical ecosystem: mapping the risk.* World Resources Institute. Washington DC. http://defiendelasierra.org/descargas/mining_critical_ecosystems_full.pdf (accessed July 17, 2012).

Morris, R.J. 2010. Anthropogenic impacts on tropical forest biodiversity: a network structure and ecosystem functioning perspective. *Philos. Trans. R. Soc. Lond. B. Biol. Sci.* **365** (**1558**): 3709-3718. Doi: 10.1098/rstb.2010.0273. http://www.ncbi.nlm.nih.gov/pmc/articles/PMC2982004/ (accessed January 23, 2013).

Muchaal, P.K. and G. Nganjuh. 1999. Impact of village hunting on wildlife population in the Western Dja Reserve, Cameroon. *Conservation Biology* **13** (**2**): 385-396.

Nijboer, J. and M. Damen. 2000. European Studbook no. 4. Crowned Pigeons *Goura cristata*, *Goura victoria* and *Goura scheepmakeri*. Rotterdam Zoo. http://www.rotterdamzoo.nl/import/assetmanager/4/6464/EEP%20Crowned%20pigeon%20n (accessed January 20, 2012).

Noss, Andrew. 2000. Cable snares and nets in the Central African Republic. **In**: *Hunting for sustainable in tropical forest*, eds. J.G. Robinson and E.L. Bennett. 282-304. New York. Columbia University Press.

Noss, A.J and B.S. Hewlett. 2001. The context of female hunting in central Africa. *American Anthropologist New Series* **103** (**4**): 1024 – 1040.

Notanubun, P.H. 2002. Tingkat kesukaan burung dara mahkota Victoria (*Goura victoria*) terhadap beberapa jenis pakan (Skripsi sarjana) Manokwari, Papua Barat: Universitas Negeri Papua.

Obizinki, K., R. Andriani, H. Kamarudin and S. Andrianto. 2012. Environmental and social impact of oil palm plantation and their implications for bio fuel production in Indonesia. Ecology and Society 17 (1): 1-25. http://dx.doi.org/10.5751/ES-04775-70125 (accessed October 31, 2012).

O'Brien, T.G., N.L. Winarni, F.M. Saanin, M.F. Kinnaird, P. Jepson. 1998. Distribution and conservation status of Bornean Peacock-pheasant *Polypectron schleiermacheri* in Central Kalimantan, Indonesia. *Bird Conservation International* **8** (**4**): 373-385 DOI 10.1017/S095ß270900002136.

O'Brien, T.G. and M.F. Kinnaird. 1996. Changing populations of birds and mammals in North Sulawesi. *Oryx* **30** (**2**):150-156

O'Brien, T.G. and M.F. Kinnaird. 2000. Differential vulnerability of large birds and mammals to hunting in North Sulawesi, Indonesia and the outlook for the future. **In**: *Hunting for Sustainability in Tropical Forests*, eds. J.G. Robinson and E.L. Bennett 199-213. New York. Columbia University Press.

Ohl-Schacherer, J., G.H. Shepard Jr, H. Kaplan, C.A. Peres, T Levi and D.W. Yu. 2007. The sustainability of subsistence hunting by Matsigenka native communities in Manu National Park, Peru. *Conservation Biology* **21** (**5**): 1174-1185. DOI: 10.1111/j.1523-1739.2007.00759.x

Padmanaba, M., M. Boissiére, Ermayanti, H. Sumatri and R. Achdiawan. 2012. Pandangan tentang perencanaan kolaboratif tata ruang wilayah di Kabupaten Mamberamo Raya, Papua, Indonesia. Studi kasus di Burmeso, Kwerba, Metaweja, Papasena dan Yoke (*Collaborative outlook of spatial planning of Mamberamo Raya Regency, Papua, Indonesia-case study in Burmeso, Kwerba, Metaweja, Papasena and Yoke)*). Center for International Forest Research. Bogor. Indonesia. Pp 82. http://www.cifor.org/mla/download/publication/Mamberamo_id_web.pdf (accessed October 4, 2012)

Pailler, S., J.E. Wagner, J.G. McPeak and D.W. Floyd. 2009. Identifying conservations opportunities among Malinké bushmeat hunters of Guinea, West Africa. *Human Ecology* **37**: 761-774.

Pangau-Adam, M. and R. Noske. 2010. Wildlife hunting and bird trade in Northern Papua (Irian Jaya), Indonesia. **In**: *Ethno-Ornithology, Birds, indigenous peoples, Culture and society*, eds. S. Tideman and A. Gosler. 73 – 85. Washington DC. Earthscan.

Pangau-Adam, M., R. Noske. and M. Muehlenberg 2012. Wildlmeat or bushmeat? subsistence hunting and commercial harvesting in Papua (West New Guinea) Indonesia. *Human Ecology*. DOI: 10.1007/S110745-012-9492-5. http://link.springer.com/content/pdf/10.1007%2Fs10745-012-9492-5 (accessed December 12, 2012)

Pangkali, L. 2011. Selamatkan manusia dan hutan Papua-suatu kajian pada studi kasus masyarakat adat di Unurumguay dan Kaureh, Kabpaten Jayapura. Jayapura Papua: PT PPMA Papua. (*in prep.*)

Pattiselanno, F. and G. Mentansan. 2010. The practice of traditional wisdom in wildlife hunting by Maybrat Etnic group to support wildlife sustainability in Sorong Selatan Regency. *Makara Sosial Humniora* **14** (**2**): 75 – 82. http://journal.ui.ac.id/index.php/humanities/article/viewFile/664/633 (accessed April 9, 2012).

Pattiselanno, F. 2008. Man-wildlife interaction: understanding the concept of conservation ethics in Papua. *Tigerpaper* **35** (**4**): 10-12 http://www.fao.org/docrep/012/ak855e/ak855e00.pdf (accessed April 9, 2012).

Pattiselano, F. 2007. Cuscus (Phalangeridae) hunting by Napan communities at Ratewi Island, Nabire, Papua. *Biodiversitas* **8** (**4**): 274-278 http://biodiversitas.mipa.uns.ac.id/D/D0804/D080406.pdf (accessed April 14, 2012)

Pattiselano, F. 2006. The wildlife hunting in Papua. *Biota* **11** (1):59-61. http://www.papuaweb.org/dlib/jr/pattiselanno/2006.pdf (accessed April 4, 2012).

Pattiselanno, F. 2003. The wildlife value: example from West Papua, Indonesia. *Tigerpaper* **30** (1):27-29. http://www.fao.org/docrep/012/ak878e/ak878e00.pdf (accessed April 9, 2012).

Pattiselano, F. and A. Arobaya. 2013. Indigenous people and nature conservation: opinion. http://www2.thejakartapost.com/news/2013/01/05/indigenous-people-and-nature-conservation.html. (accessed January 7, 2013)

Peckover, W.S. and L.W.C. Filewood. 1976. *Birds of New Guinea and Tropical Australia*. Cornel Universiy Press.

Peres, C.A. 2000. Evaluating the impact and sustainability of subsistence hunting of multiple Amazonian forest site. **In:** *Hunting for Sustainability in Tropical Forests*, eds. J.G. Robinson and E.L. Bennett . 31-56. New York. Columbia University Press.

Perrin, C. (Ed). 2009. The Enchyclopedia of Birds. New York. Oxford University Press.

Peroni, N. and N. Hanazaki. 2002. Current and lost diversity of cultivated varieties, especially cassava under swedden cultivation systems in the Brazilian Atlantic Forest. *Agriculture, Ecosystem and Environment* **92**:171-183

Petocs, R.G. 1978. *Conservation and Development in Irian Jaya. A Strategy for Rational Resource Utilization.* Leiden. E.J. Brill.

Philips, A. 2001. Mining and protected area. *Mining, Minerals and Sustainable Development* **62**: 1-62. http://www.miningnorth.com/docs/Mining%20and%20Protected%20Areas.pdf (accessed October 4, 2012).

Pratt, T. K. 1982. Biogeography of Birds in New Guinea. In: *Monographiae Biologicae* Volume **42,** eds. J. Illies and F.R.G. Schlitz. In: *Biogeography and Ecology of New Guinea* Volume **1**, ed. J.L. Gressitt. 815-833. London. Dr. W. Junk Publishers. The Hague.

Profauna, 2012. Parrots smuggling from Indonesia to Philipine http://www.profauna.org/content/pirated_parrots.html, (accessed May 6, 2012).

Purnama, S. and M. Indrawan. 2010. Entrapment of wetland birds: local customs and Methods of hunting in Krangkeng, Indramayu, Central Java. **In:**

Ethno- Ornithology, Birds, indigenous peoples, Culture and society. S. Tideman and A. Gosler. 67-72. Washington DC. Earthscan.

Purwaningsih and R. Yusuf. 2008. Analisa vegetasi hutan pegunungan di Taman Nasional Gunung Ciremai, Majalengka, Jawa Barat (*The mountain rain forest vegetation analysis in Ciremai Mountain national Park, Majalengka, West Java*). Jurnal Biology Indonesia 4 (5): 385-399

Rao, M., Than Myint, Than Zaw and Saw Hun. 2005. Hunting patterns in tropical forests adjoining the Hkakaborazi National Park, North Myanmar. *Oryx* **39** (**3**): 292-300.

Rand, A.L. and E.T. Gillard. 1967. *Handbook of New Guinea Birds.* Weidenfield and Nicholson. London. Pp. 187-189

Randrianandrianina; F.H., P.A. Racey and R.B. Jenkins. 2010. Hunting and consumption of mammals and birds by people in urban areas of Western Madagascar. *Oryx* **44** (**3**): 411-415

Roberts, R.G., T.F. Flanery, L.K. Ayliffe, H. Yoshida, J.M. Olley, G.J. Prideaux, G.M. Laslett, A. Baynes, M.A. Smith, R. Jones, and B.L. Smith. 2001. The last Australian megafauna: new ages indicate continent-wide extinction about 46.000 years ago. *Science* **292** (**5523**):1888-1892 http://web.ebscohost.com/ehost/detail?sid=914a9910-8aa1-496d-9f19-2a221ab35cfe%40sessionmgr10&vid=5&hid=12&bdata=JnNpdGU9ZW Whvc3QtbGl2ZQ%3d%3d#db=pbh&AN=4771056 (accessed January 16, 2013).

Robinson, J.G and K.H. Redford. 1994. Measuring the sustainability on hunting in tropical forests. *Oryx* **28** (**4**): 249-256.

Robinson, J.G. 2000. Appendix: Calculating maximum sustainable harvests and percentage offtakes **In**: *Hunting for Sustainability in Tropical Forests,* eds. J.G. Robinson and E.L. Bennett. 521-524. New York. Columbia University Press.

Robinson, J.G., K.H. Redford and E.L. Bennett. 1999. Wildlife harvest in logged tropical forest. *Science* **284** (**5414**): 595-561

Rosenbaum, H., T.G. O'Brien; M.F. Kinnaird and J. Supriatna. 1998. Population densities of Sulawesi crested black macaques (*Macaca nigra*) on Bacan and Sulawesi, Indonesia. Effect of habitat disturbance and hunting. *American Journal of Primatology* **44**: 89-106.

Richards, P.W. 1996. *The tropical rain forest an ecological study.* 2nd Edition. Cambridge. Cambridge University Press.

Richard, S.J. and S. Suryadi. (Eds). 2002. A biodiversity assessment of Yongsu-Cyclops mountain and the southern Mamberamo Basin, Papua,

Indonesia. Rapid Assessment Program (RAP). Bulletin of Biological Assessment No. 25. Washington DC. Conservation International.

Riley, J. 2002. Mammals on the Sangihe and Talaud Island, Indonesia, and the impact of hunting and habitat loss. *Oryx* **36** (**3**): 288-296

Saaroni, Y and H. Simbolon. 1998. Birds and mammals in Supiori island and North Biak Nature Reserve, Irian Jaya in: H. Simbolon (Ed). Irian Jaya: Plants and animals research miscellany of Biak-Supiori and Yapen Islands. Bogor, Indonesia: Puslitbang LIPI

Sada, J.J. 2005. Sistem perburuan burung Mambruk Victoria (*Goura victoria beccarii*) oleh masyarakat kampung Awaso dan Somiangga Distrik Waropen Bawah Kabupaten Waropen (Skripsi sarjana), Manokwari, Papua Barat: Universitas Negeri Papua.

Saleh, Ch. 2005. Perdagangan illegal hidupan liar di Indonesia: potret penegakan hukum. Makalah dalam Prosidings Judicial Workshop Penegakan Hukum Atas Perlindungan Satwa Liar. Tobing, H., F. Hanif dan R. Setyaningrum (Editors). Kerjasama WWF Indonesia-TRAFFIC South Asia- Mahkamah Agung Republik Indonesia. Cibodas 27-29 September 2005. 11 pp. http://rafflesia.wwf.or.id/library/admin/attchment/books/prosidings%20versi%20pdf.pdf (accessed January 17, 2013).

Samuelson, R. 2008. West Papua is Indonesia's oil palm target. http://www.infopapua.org?artman/exee/view.cgi?archieve=318&num=1747 (accessed October 25, 2011)

Sanggenafa, N. 1992. Masyarakat Waropen di pantai Timur Teluk Cenderawasih **in**: *Irian Jaya membangun masyarakat majemuk*. Seri Etnografi Indonesia 5, ed. Koentjaraningrat. 214-229. Jakarta. Penerbit Djambatan.

Seiler, A. 2001. Ecological effect of roads- a review. Department of Conservation Biology-Swedish University of Agriculture Science. *Introductory Research Essay* no. **9**. Upsala. Swedish. http://idd00s4z.eresmas.net/doc/transp/ecoeffectsonroads.pdf (accessed October 4, 2012)

Sembiring, S.N.. 2005. Implementasi peraturan terkait perlindungan hidupan liar. Paper **in**: Prosidings Judicial Workshop Penegakan Hukum Atas Perlindungan Satwa Liar. Tobing, H., F. Hanif dan R. Setyaningrum (Editors). Kerjasama WWF Indonesia-TRAFFIC South Asia- Mahkamah Agung Republik Indonesia. Cibodas 27-29 September 2005. 4 pp.

http://rafflesia.wwf.or.id/library/admin/attchment/books/prosidings%20v
ersi%20pdf.pdf (accessed January 17, 2013).

Setio, P., H. Alhamid and Y.O. Leikitoo. 1996. Pola perkembangbiakan burung
dara mahkotaViktoria *Goura victoria Bull. Penelitian Kehutanan* **2**: 1-9
http://agris.fao.org/agris-
search/search/display.do?f=2000/ID/ID00001.xml;ID2000000389
(accessed January 15, 2012).

Setio, P. and Y.O. Leikitoo. 2000. Peranan burung kasuari dan satwa liar lainnya
dalam regenerasi hutan di Irian Jaya. *Matoa* **7**:51-57.

Shaw, D.E. 1969. Conservational Ordinances in Papua and New Guinea.
Biological Conservation **2** (**1**): 50-53.

Sheil, D., R.K. Puri, I. Basuki, M. Van Heist, M. Wan, N. Liswanti, Rukmiyati,
M.A. Sardjono, I. Samsoedin, K. Sidiyasa, Chrisdanini, E. Permana,
E.M. Angi, F. Gatzweiler, B. Johnson dan A. Wijaya. 2004.
Mengeksplorasi keanekaragaman hayati, lingkungan dan pandangan
masyarakat lokal mengenai berbagai lanskap hutan; metode-metode
penilaian lanskap secara multidisipliner (*Exploring biological diversity,
environment and local people's perspectives: multidisciplinary
landscape assessment methods*). Bogor. CIFOR.

Siahaan, L. 2006. Keragaman genetic Cytochrome B pada burung mambruk
(*Goura* sp) (skripsi sarjana) Bogor, Indonesia: Institut Pertanian Bogor..

Sillitoe, P. 2002. Always been farmer-forager? Hunting and gathering in the
Papua New Guinea Highland. *Anthropological Forum* **12** (**1**): 45 -76.

Silva, J.L. and S.D. Strahl. 1991. Human impact on population of Chachalacas,
Guans and Currashaws (Galliformes: Cracidae) in Venezuela. In:
Neotropical Wildlife Use and Conservation. Eds. Robinson, J.G. and
K.H. Redford, 36-52. Chicago. University of Chicago Press.

Smith, D.A. 2005. Garden game: shifting cultivation, indigenous hunting and
wildlife ecology in western Panama. *Human Ecology* **33**: 505-537

Smith, D.A. 2010. The harvest of rain-forest birds by indigenous communities in
Panama. *The Geographical Review* **100** (**2**): 187-203.

Smolker, R., B. Tokar, A. Petermann and E. Hernandes. 2007. Devastated lands,
displaces people agrofuels cost in Indonesia, Malaysia and Papua New
Guinea in the real costs of agrofuels: foods, forests and climate.
http://www.pasificecologist.org/archieve/17/pe17-biofuels-devastate-se-
asia.pdf (accessed October 25, 2011).

Snow, D. S. 1981. Tropical frugivorous birds and their food plants: a world survey. *Biotropica* **13** (**1**): 1 - 14

Sodhi, N.S., Lian Pin Koh and B.W. Brook. 2006. Southeast Asian birds in peril. *The Auk* **123** (**1**): 275-277

Sodhi, N.S., R. Butler, W.F. Laurance and L. Gibson. 2011[a]. Conservation successes at micro, meso- and macroscales. TRENDS Ecology and Evolution 26 (11): 585-594 .DOI 10.1016/j.tree.2011.07.002

Sodhi, N.S., R. Butler and P.H. Raven. 2011[b]. Bottom-up Conservation (*Commentary*). *Biotropica* **43** (**5**): 521-523. DOI: 10.1111/j.1744-7429.2011.00793.x

Statterfields, A.K., M.J. Crosby, A.J. Long and D.C. Wege. 1998. Endemic Bird Areas of the world: Priorities for Biodiversity Conservation. Cambridge: BirdLife International.

Sujatnika. P. Jepson, T.R. Suhartono, M.J. Cosby and A. Mardiastuti. 1995. Melestarikan keanekaragaman hayati Indonesia: pendekatan Daerah burung endemik (*Conservation biodiversity of Indonesia: approaches of endemic bird area*). Jakata. Directorate of Forest Protection and Nature Conservation/BirdLife International Indonesia Programme.

Supriatna, J. 1997. (Ed). Laporan Akhir: Lokakarya penentuan prioritas konservasi keanekaragamam hayati Indonesia (*Final report: The Irian Jaya Biodiversity Conservation priority-setting workshop*). Washington DC. Conservation International.

Supriatna, J. 2008. Melestarikan alam Indonesia. Jakarta. Yayasan Obor Indonesia.

Suprayitno, A. 2007. Pendugaan model pertumbuhan populasi dan daya dukung habitat wallaby lincah (*Mocropus agilis papuanus*, Pieters and Doria 1875) di Taman Nasional Wasur. (Master thesis) Bogor, Indonesia: Institut Pertanian Bogor.

Suryadi, S., A. Wijayanto and J.B. Cannon. 2007. Conservation laws, regulation, and Legislation in Indonesia, with special reference to Papua. **In**: *The Ecology of Papua part Two*, eds. A.J. Marshall and B.M. Beehler. 1276-1310. Singapore. Periplus Editions.

Thornton, D.H., L.C. Branch and M.E. Sunquist. 2011. Response of large Galliforms and tinamous (Cracidae, Phasianidae, Tinamidae) to habitat loss and fragmentation in northern Guatemala. *Oryx* **46** (**4**): 567-576

Tien Ming Lee, N.S. Sodhi and D.M. Prawiradilaga. 2009. Determinants of local people'attitude towards conservation and the consequential effect on illegalresources harvesting in the protected areas of Sulawesi (Indonesia). *Environmental Conservation* **36** (**2**): 157-170. DOI: 10.1017/S0376892909990178

Tribisono, H. 2002. Tingkah laku makan burung dara mahkota Cristata (*Goura cristata*) pada lingkungan penangkaran di Taman Burung dan Taman Anggrek Kabupaten Biak Numfor. (Skripsi sarjana) Manokwari, Indonesia: Universitas Negeri Papua.

Torobi, P.M. 2005. Inventarisasi jenis-jenis burung peliharaan yang dilindungi di Sentani Jayapura. (Skripsi sarjana) Jayapura, Papua: Universitas Cenderawasih.

Towsend, W.R. 2000. The sustainability of subsistence hunting by Sirionó Indians of Bolivia. **In**: *Hunting for sustainable in tropical forest*, eds. J.G. Robinson and E.L. Bennett. 267-281. New York. Columbia University Press.

Van Vliet, N. and R. Nasi. 2008. Hunting for livelihood in northeast Gabon: patterns, evolution and sustainability. Ecology and Society 13 (2): 33 http://www.ecologyandsociety.org/vol13/iss2/art337 (accessed October 23, 2012).

Wadley, R.L., C.J.P. Colfer and I.G. Hood. 1997. Hunting primate and managing forests: the case of Iban forest farmers in Indonesia Borneo. *Human Ecology* **25** (**2**):243-271.

Wadley, R.L. and C.J.P. Colfer. 2004. Sacred forest, hunting and conservation in West Kalimantan, Indonesia. *Human Ecology* **22** (**3**): 313-337

Walker, J.S., A.L. Cahill and S.J. Marsden. 2005. Factors influencing nest site occupancy and low productive output in critically endangered Yellow-crested cockatoo cacatua sulphurea on Sumba, Indonesia. *Bird Conservation International* **15**: 347-359. DOI 101017/s 0959270905000638.

Wamebu, N. 2000. Pemetaan partisipatif multipihak: wilayah adat Nambluong di Kabupaten Jayapura-Papua. http://www.jkpp.org/downloads/04.%20Papua.pdf (accessed January 17, 2013).

Waltert, M., Lien, K. Faber and M. Mühlenberg.2002. Further decline of threatened primates in the Korup Project Area, south-west Cameroon. *Oryx* **36** (**3**): 257-265

Waltert, M., C. Sheifert, G. Radhl and B. Hope-Dominik. 2010. Population size and habitat of the White-breasted Guineafowl Agelastes meleagrides in the Täi region, Côte d'Ivoire. *Bird Conservation International* **20**: 74-83. DOI 10.1017/s0959270909990189

Wilkie, D. S., 1989. Impact of roadside agriculture on subsistence hunting in the Ituri forest of northeastern Zaire. *American Journal of Physical Anthropology* **78**: 485-494

Wilson, W.L. and A.D. Johns. 1982. Diversity and abundance of selected animal species in undisturbed forest, selectively logged forest and plantations in East Kalimantan, Indonesia. *Biology Conservation* **24**: 205-218.

Winarni, N.L. and M. Johns. 2012. Effect of anthropogenic disturbance on the abundance and habitat occupancy of two endemic hornbill species in Buton Island, Sulawesi. *Bird Conservation International* **22** (**2**): 222-233 DOI: 10.1017/S0959270911000141

Winarni, N.L., T.G. O'Brien and J.P. Carol. 2009. Movement, distribution and abundance of great Argus Pheasants (*Argusianus argus*) in a Sumatera rainforest. *The Auk* **126** (**2**): 341-350

World Wildlife Fund Global. 2012. Freeport Mine: in the wake of hungry giant. http://wwf.panda.org/what_we_do/where_we_work/new_guinea_forests/ problems_forersts_new_guinea/mining_new_guinea/papua_freeport_mi ne/ (accessed October 31, 2012).

Yaap, B., M.J. Struebig, G. Paoli and Lian Pin Koh. 2010. Mitigating the biodiversity impacts of oil palms development. *Perspective in Agriculture, Veterinary Science, Nutrition and natural Resources* **5** (**019**) DOI 10.1017/PAVSN NR20105019 http://www.hcvnetwork.org/resources/folder.2006-09-29.6584228415/Yaap_etal_2010_CABReviews_5.pdf (accessed January 18, 2012)

Appendix 1. Table of tree species in forest of Buare site

No.	Scientfic name	Family	DeR (%)	FR (%)	DoR (%)	IVP (%)
1	*Ficus subulata*	Moraceac	0.56	0,57	0,108	1,24
2	***Myristica sp2***	**Myristicaceae**	6.11	6,32	2,098	14,5
3	*Arthocarpus vriseanus*	Moraceae	0.56	0,57	0,256	1,39
4	*Campnosperma macrophyla*	Anacardiaceae	2.78	2,3	2,386	7,46
5	*Canarium maluense*	Burseraceae	2.22	2,3	1,682	6,2
6	*Lophopetalum javanicum* (Zoll.)Turcz.	Hammelidae	0.56	0,57	0,576	1,71
7	*Garcinia holrungii* Lautrb	Cluciaceae	0.56	0,57	0,256	1,39
8	*Calophyllum* sp.	Guttiferae	0,56	0,57	0,256	1,39
9	*Garcinia celebica* Linn	Cluciaceae	0,56	0,57	1,296	2,43
10	*Garcinia dulcis* (Roxb.) Kurz.	Cluciaceae	2,22	2,3	1,561	6,08
11	***Terminalia microcarpa* Decne**	**Combretaceae**	2,78	2,3	4,649	9,73
12	*Dillenia auriculata* Mart.	Dilleniaceae	0,56	0,57	0,064	1,19
13	***Dracontomelum edule* Merr**	**Anacardiaceae**	1,67	1,72	5,136	8,53
14	*Horsfieldia batjanica*	Myristicaceae	1,11	1,15	1,273	3,53
15	*Gnetum gnemon*	Gnetaceae	1,11	1,15	0,832	3,09
16	*Terminalia cannliculata*	Combretaceae	0,56	0,57	0,077	1,21
17	***Pimeliodndron amboinicum* Hassk.**	**Euphorbiaceae**	11,7	11,5	15,37	38,5
18	***Intsia* spp**	**Fabaceae**	6,67	6,9	6	19,6
19	*Lithocarpus celebicus* (Miq.) Rehder	Fagaceae	0,56	0,57	0,784	1,91
20	*Gonocaryum littorale*	Icacynaceae	1,11	1,15	1,152	3,41
21	*Litsea forstenii* (Bl.) Boerl	Lauraceae	0,56	0,57	0,016	1,15
22	*Tetrameles nudiflora*	Tetramelaceae	1,67	1,72	0,621	4,01
23	*Cryptocarya infectoria*(B.l)	Lauraceae	2,78	2,87	0,554	6,2
24	*Planchonia papuana* Kunth.	Sapotaceae	2,78	2,3	1,685	6,76
25	*Litsea firma* Hook.f	Lauraceae	0,56	0,57	0,576	1,71
26	*Cryptocarya multipaniculata*	Lauraceae	1,11	1,15	0,953	3,21
27	*Gluta renghas*	Anacardiaceae	0,56	0,57	0,064	1,19
28	*Mangifera* spp	Anacardiaceae	0,56	0,57	0,077	1,21
29	*Aglaia odorata* Kour.	Cyperacea	0,56	0,57	1,6	2,73
30	*Dysoxylum arborescens* Miq.	Meliaceae	0,56	0,57	1,024	2,15
31	*Homalium foetidum*	Flacourtiaceae	0,56	0,57	0,576	1,71
32	*Horsfieldia silvestris* Warb	Myrtaceae	2,22	2,3	2,509	7,03
33	*Ficus adenosperma* Miq	Moraceae	1,67	1,72	1,337	4,73
34	***Eugenia anomala* Lauth**	**Myrtaceae**	6,67	6,32	12,65	25,6
35	*Elaiocarpus spaherius* K.Schal	Elaiocarpaceae	0,56	0,57	0,502	1,63
36	*Drypetes macrophylla* Bl	Euphorbiaceae	0,56	0,57	0,256	1,39
37	*Timonius timon*	Rubiaceae	0,56	0,57	0,576	1,71
38	*Anthocephalus cadamba* Mig	Rubiaceae	0,56	0,57	0,697	1,83

No	Species	Family			
39	*Pometia pinata* **J.R.&G.Forst.**	**Sapindaceae**	5,56	5,75	4,566	15,9
40	*Palaquium ridleyi* K.& G.	Sapotaceae	2,78	2,87	3,376	9,03
41	*Palaquim amboinensis*	Sapotaceae	2,78	2,87	1,29	6,94
42	*Paraserianthes falcataria* Baker	Leguminoceae	1,11	1,15	0,216	2,48
43	*Myristica papua* Mkgf	Myristicaceae	1,67	1,72	1,76	5,15
44	*Syzygium* **sp.**	**Myrtaceae**	7,22	6,9	9,44	23,6
45	*Teismanniodendron ahenianum* (Merr.) Bakh.	Verbenaceae	0,56	0,57	1,296	2,43
46	*Vitex cofassus* Reinw.	Verbenaceae	0,56	0,57	0,655	1,79
47	*Podocarpus blumei*	Podocarpaceae	0,56	0,57	0,092	1,22
48	*Macaranga gigantea*	Euphorbiaceae	1,11	1,15	0,346	2,61
49	*Sterculia parkinsonii* F.V.M	Sterculiaceae	0,56	0,57	0,256	1,39
50	*Polyathia subcordata* Bl	Annonaceae	0,56	0,57	0,144	1,27
51	*Horsfildea silvestris*	Myristicaceae	0,56	0,57	0,144	1,27
52	*Conandrium polyanthum* Miq	Myrsinaceae	0,56	0,57	0,576	1,71
53	*Celtis* spp	Ulmaceae	1,11	1,15	1,424	3,68
54	*Octomeles sumatrana*	Datiscaceae	0,56	0,57	0,144	1,27
55	*Caseoria glabra*	Flacourtiaceae	0,56	0,57	0,256	1,39
56	*Haplolobus lanceolatus* HJL	Burseraceae	0,56	0,57	0,256	1,39
57	*Mastixiodendron* sp	Rubiaceae	1,11	1,15	1,424	3,68
58	*Manielkara kanosinensis* HJL et BM	Sapotaceae	0,56	0,57	0,256	1,39
	58		**100**	**100**	**100**	**300**

Appendix 2. Table of tree species in forest of Supiori site

No.	Nama Jenis	Family	DeR (%)	FR (%)	DoR (%)	IVP (%)
1	*Actinodaphne nitida* Jesch	Lauraceae	0,727	0,48	0,11	1,318
2	*Aglaia argentata* Bl	Meliaceae	3,273	3,837	3	10,11
3	*Aglaia specibs*	Meliaceae	0,364	0,48	0,37	1,216
4	*Beiesmeldia bulata* Allen	Lauraceae	0,364	0,48	0,17	1,009
5	***Blumeodendron* sp**	**Euphorbiaceae**	5,455	3,837	2,37	11,66
6	***Callophyllum inophyllum* Linn**	**Guttiferae**	6,909	6,715	5,24	18,86
7	*Campnosperma* sp	Anacardiaceae	0,364	0,48	0,45	1,295
8	***Canarium indicum* L.**	**Burseraceae**	5,818	6,235	12,5	24,58
9	*Caralia brachinata* Merr	Rhizophoraceae	1,818	1,918	0,68	4,419
10	*Celtis laetifolia* Planch	Ulmaceae	0,364	0,48	0,45	1,295
11	*Corinocarpus* sp	Corynocarpaceae	0,364	0,48	0,04	0,885
12	*Cryptocarya palmerensis* Allen	Lauraceae	1,455	1,918	0,68	4,055
13	*Dilenia alata* Mart	Dilleniaceae	0,727	0,48	0,11	1,318
14	*Drypetes globossa*	Euphorbiaceae	1,455	1,918	1,27	4,641
15	*Elaeocarpus spahaericus* K.Schl	Elaiocarpaceae	0,364	0,48	0,84	1,683
16	*Endospermum moluccanum* Becc	Euphorbiaceae	0,727	0,48	0,19	1,393
17	***Eugenia anomala* Lauth**	**Myrtaceae**	8,727	6,715	8,98	24,42
18	*Eugenia versteghi* Lauth	Myrtaceae	0,364	0,48	0,09	0,937
19	*Fagraea* sp	Loganiceae	0,364	0,48	0,17	1,009
20	*Ficus altissima*	Moraceae	0,364	0,48	2,33	3,176
21	*Ficus variegata* BL	Moraceae	0,364	0,48	0,84	1,683
22	*Flindercia amboinensis* Poir	Rutaceae	1,818	1,918	3,3	7,039
23	*Garcinia dulcis*	Guttiferae	0,364	0,48	0,04	0,885
24	*Gonocaryum piriform* Scheff	Icacynaceae	1,455	1,918	0,23	3,599
25	*Gymnacranthera paniculata* Warb	Myristicacea	0,727	0,959	0,43	2,119
26	*Haplolobus floribundus* HJL	Burseraceae	1,455	1,918	1,67	5,04
27	*Haplolobus lanceolatus* HJL	Burseraceae	0,364	0,48	0,08	0,925
28	*Heritiera* sp	Malvaceae	0,364	0,48	0,37	1,216
29	*Hoemalium foetidum* Bth	Flacourtiaceae	0,364	0,48	0,66	1,507
30	***Homonoia javanensis* M.A.**	**Euphorbiaceae**	10,91	6,715	1,62	19,24
31	*Hopea iriana* Slooth	Dipterocarpaceae	0,727	0,959	0,42	2,11
32	*Horsfieldia silvestris* Warb	Myristicacea	0,364	0,48	0,33	1,168
33	*Intsia palembanica* Miq	Fabaceae	2,182	2,878	3,28	8,34
34	***Lepiniopsis ternatensis* Vall**	**Apocynaceae**	4,364	4,796	2,16	11,32
35	*Litsea tuberculata*	Lauraceae	0,364	0,48	0,07	0,913
36	*Macaranga mappa* BL	Euphorbiaceae	0,727	0,959	0,14	1,83
37	*Maniltoa brownoides* Harms	Sapotaceae	0,727	0,959	0,14	1,83

38	*Melastoma* sp	Melastomaceae	0,727	0,959	0,47	2,153
39	*Myristica fatua*	Myristicaceae	1,455	1,439	0,78	3,671
40	*Myristica papuana* Mkgf	Myristicaceae	1,455	1,439	0,56	3,452
41	***Myristica tubiflora* BL**	**Myristicaceae**	2,909	2,878	0,53	6,319
42	**Palaquium lobianumBurck**	**sapotaceae**	3,273	2,878	12,5	18,66
43	*Penthapalangium pachycarpum* Ac.sm	Guttiferae	0,364	0,48	4,15	4,99
44	***Pimeliodendron amboinicum* Hassk**	**Euphorbiaceae**	6,182	5,755	5,35	17,29
45	***Planchonella anteridifera***	**Sapotaceae**	5,818	4,796	11,9	22,55
46	*Planchonella odorata* Piere	Sapotaceae	1,818	1,439	1,12	4,375
47	*Podocarpus amara* BL	Podocarpaceae	0,727	0,959	0,3	1,987
48	*Podocarpus blumei* Linn	Podocarpaceae	0,364	0,48	0,04	0,885
49	*Pometia acuminata* Radlk	Sapindaceae	0,727	0,959	0,54	2,226
50	*Pometia pinnata* Forst	Sapindaceae	0,364	0,48	0,04	0,885
51	*Pterygota horsfieldii* Kosterms	Sterculiaceae	1,818	2,398	2,93	7,141
52	*Pygeum parviflorum* TB	Rosaceae	0,364	0,48	0,26	1,102
53	*Reindwardtiodendron celebicum* Kds	Meliaceae	0,364	0,48	0,37	1,216
54	*Sloanea pullei* Ac.sm	Elaiocarpaceae	0,727	0,959	0,6	2,288
55	*Terminalia canaliculata*	Combretaceae	0,364	0,48	0,09	0,937
56	*Teysmaniodendron bogoriense* Kds	Verbenaceae	1,455	1,918	1,29	4,664
57	*Uranda brasii* How	Icacynaceae	0,727	0,959	0,28	1,963
	57		**100**	**98,8**	**100**	**298,8**

Appendix 3. Table of tree pecies in forest of Unurumguay site

No.	Scientific name	Family	DeR (%)	FR (%)	DoR (%)	IVP (%)
1	*Alstonia scholaris*	Apocynaceae	1,130	1,190	0,184	2,504
2	*Arthocarpus altilis*	Moraceae	1,130	1,190	0,356	2,676
3	*Callophyllum inophylum*	Guttiferae	1,695	1,190	0,422	3,308
4	*Campnosperma auriculatum*	Anacardiaceae	6,215	6,548	4,208	16,97
5	*Cananga odorata*	Annonaceae	1,130	1,190	0,307	2,628
6	*Cananga spp*	Annonaceae	0,565	0,595	0,614	1,774
7	*Canarium moluccensis*	Burseraceae	2,260	1,786	2,542	6,588
8	*Canarium spp*	Burseraceae	0,565	0,595	0,743	1,903
9	*Celtis latifolia*	Ulmaceae	0,565	0,595	0,154	1,314
10	*Dracontomelum edule* Merr	Anacardiaceae	3,955	4,167	9,605	17,73
11	*Dyllenia grandiflora*	Dilleniaceae	0,565	0,595	0,743	1,903
12	*Endospermum diadenum*	Euphorbiaceae	1,695	1,190	1,311	4,196
13	*Eugenia anomala*	Myrtaceae	0,565	0,595	0,068	1,228
14	*Eugenia spp*	Myrtaceae	0,565	0,595	0,068	1,228
15	*Ficus hispida*	Moraceae	1,130	1,190	2,572	4,892
16	*Ficus pungens*	Moraceae	2,825	2,976	1,468	7,269
17	*Garcinia dulcis*	Cluciaceae	1,695	1,786	0,528	4,009
18	*Gnetum gnemon*	Gnetaceae	0,565	0,595	6,824	7,984
19	*Intsia spp*	Fabaceae	5,650	5,952	6,378	17,98
20	*Litsea tuberculata*	Lauraceae	0,565	0,595	0,154	1,314
21	*Macaranga gigantea*	Euphorbiaceae	0,565	0,595	1,706	2,866
22	*Macaranga mappa*	Euphorbiaceae	0,565	0,595	0,614	1,774
23	*Manielkara spp*	Sapotaceae	0,565	0,595	0,273	1,433
24	*Mastixiodendron spp*	Rubiaceae	1,130	1,190	0,682	3,003
25	*Myristica fatua*	Myristicaceae	2,825	2,976	2,286	8,087
26	*Myristica papuana* Kunth	Myristicaceae	10,169	10,714	6,054	26,94
27	*Octomeles sumatrana*	Datiscaceae	1,130	0,595	1,706	3,431
28	*Palaquium amboinensis*	Sapotaceae	4,520	4,167	3,61	12,3
29	*Pimeliodendron amboinicum* Hassk	Euphorbiaceae	11,299	11,310	17,98	40,59
30	*Pometia spp*	Sapindaceae	12,994	13,690	12,09	38,78
31	*Pterygotha horsfildea*	Sterculiaceae	8,475	7,738	7,622	23,83
32	*Spondias dulcis*	Anacardiaceae	0,565	0,595	0,273	1,433
33	*Syzigium spp*	Myrtaceae	1,130	1,190	0,768	3,088
34	*Terminalia catappa*	Combretaceae	1,695	1,786	0,296	3,777
35	*Terminalia complanata*	Combretaceae	0,565	0,595	0,115	1,276
36	*Tetrameles nudiflora*	Tetramelaceae	0,565	0,595	0,427	1,587
37	*Teysmaniodendron bogoriens*	Verbenaceae	5,085	4,762	3,478	13,32
38	*Vitex cofassus*	Verbenaceae	0,565	0,595	0,614	1,774
39	*Vitex pubescens*	Verbenaceae	0,565	0,595	0,154	1,314
	39		100	100	100	300

Appendix 4. Table of tree species in forest of Bonggo site

No.	Scientific name	Family	DeR (%)	FR (%)	DoR (%)	IVP (%)
1	*Alstonia scholaris* R.Br.	Apocynaceae	1,316	1,316	1,481	4,113
2	*Anisotera* spp	Dipterocarpaceae	0,658	0,658	0,589	1,905
3	*Anthocephalus chinensis* (Lamk) A.Richex Walp.	Rubiaceae	0,658	0,658	0,648	1,964
4	*Arthocarpus* spp	Moraceae	0,658	0,658	0,981	2,297
5	*Callophyllum inophyllum* L.	Guttiferae	1,974	1,974	2,430	6,378
6	*Campnosperma auriculatum* (Bl) Hook.f.	Anacardiaceae	5,921	5,921	7,915	19,757
7	*Cananga* spp	Annonaceae	0,658	0,658	1,270	2,586
8	*Canarium* spp	Burseracea	4,605	4,605	7,654	16,865
9	*Celtis* sp	Ulmaceae	1,316	1,316	2,145	4,776
10	*Dillenia grandifolia* Wall	Dilleniaceae	0,658	0,658	0,981	2,297
11	*Disoxylon* sp	meliaceae	0,658	0,658	0,957	2,273
12	*Dispyros* spp	Ebenaceae	0,658	0,658	0,957	2,273
13	*Dracontomelum edule* Merr	Anacardiaceae	1,974	1,974	2,055	6,003
14	*Endospermum diadenum* (Miq.) Airy Shaw	Euphorbiaceae	5,263	5,263	6,339	16,865
15	*Ficus* spp	Moraceae	0,658	0,658	0,863	2,178
16	*Intsia bijuga* OK	Fabaceae	2,632	2,632	3,608	8,871
17	*Blumeodendron* sp	Euphorbiaceae	7,237	7,237	7,483	21,956
18	*Myristica papuana*	Myristicaceae	11,842	11,842	10,110	33,794
19	*Vitex pubescens*	Verbenaceae	0,658	0,658	0,552	1,868
20	*Tetrameles nudiflora*	Datiscaceae	0,658	0,658	0,863	2,178
21	*Manielkara kanosiensis* H.j.L. et B.M.	Sapotaceae	1,316	1,316	1,178	3,810
22	*Mastixodendron* spp	Rubiaceae	1,974	1,974	1,979	5,926
23	*Octomeles sumatrana* Miq	Datiscaceae	1,316	1,316	0,965	3,596
24	*Palaquium amboinensis*	Sapotaceae	7,237	7,237	6,309	20,782
25	*Parartocarpus* spp	Myristicaceae	1,316	1,316	0,694	3,325
26	*Paraserianthes falcataria* (L.) Nielsen	Leguminoceae	1,316	1,316	0,814	3,446
27	*Pometia* spp	Sapindaceae	10,526	10,526	9,324	30,377
28	*Pterocarpus indicus* Willd	Fabaceae	5,921	5,921	6,157	17,999
29	*Pterygota horsfieldia*	Sterculiaceae	2,632	2,632	1,944	7,207
30	*Spondias* spp	Anacardiaceae	0,658	0,658	0,552	1,868
31	*Syzygium* spp	Myrtaceae	13,158	13,158	8,934	35,250
32	*Terminalia* spp	Combretaceae	0,658	0,658	0,311	1,626
33	*Toona sureni* Merr	Meliaceae	0,658	0,658	0,311	1,626
34	*Vatica rassak* Bl.	Dipterocarpaceae	0,658	0,658	0,648	1,964
	34		**100**	**100**	**100**	**300**

Appendix 5. Plant genera were recorded as the diets of frugivorous birds (including family Columbidae) specialist and non specialist(based on Snow 1981 and Frith *et al* 1976) in each study site

Buare		Supiori	
Plant species	**Plant family**	**Plant species**	**Plant family**
Ficus subulata	Moraceae	*Canarium indicum* L.	Burseraceae
Arthocarpus vriseanus	Moraceae	*Celtis laetifolia* Planch	Ulmaceae
Canarium maluense	Burseraceae	*Cryptocarya palmerensis* Allen	Lauraceae
Celtis spp	Ulmaceae	*Drypetes globossa*	Euphorbiaceae
Cryptocarya infectoria(B.l)	Lauraceae	*Elaeocarpus spahaericus* K.Schl	Elaiocarpaceae
Drypetes macrophylla Bl	Euphorbiaceae	*Endospermum moluccanum* Becc	Euphorbiaceae
Dysoxylum arborescens Miq.	Meliaceae	***Eugenia anomala* Lauth**	**Myrtaceae**
Elaiocarpus spaherius K.Schal	Elaiocarpaceae	***Eugenia versteghi* Lauth**	**Myrtaceae**
***Eugenia anomala* Lauth**	**Myrtaceae**	*Fagraea* sp	Loganiceae
Ficus adenosperma Miq	Moraceae	*Ficus altissima*	Moraceae
Litsea firma Hook.f	Lauraceae	*Ficus variegata* BL	Moraceae
Litsea forstenii (Bl.) Boerl	Lauraceae	*Gymnacranthera paniculata* Warb	Myristicaceae
***Myristica papua* Mkgf**	**Myristicaceae**	*Litsea tuberculata*	Lauraceae
Myristica sp2	**Myristicaceae**	***Myristica fatua***	**Myristicaceae**
***Syzygium* sp.**	**Myrtaceae**	***Myristica papuana* Mkgf**	**Myristicaceae**
Terminalia cannliculata	Combretaceae	***Myristica tubiflora* BL**	**Myristicaceae**
Terminalia microcarpa Decne	Combretaceae	*Planchonella anteridifera*	Sapotaceae
Timonius timon	Rubiaceae	*Planchonella odorata* Piere	Sapotaceae
Vitex cofassus Reinw.	Verbenaceae	*Pygeum parviflorum* TB	Rosaceae
		Sloanea pullei Ac.sm	Elaiocarpaceae
		Terminalia canaliculata	Combretaceae
19 Species	**12 Family**	**21 Species**	**12 Family**

126

Appendix 5 : continued

Unurumguay		Bonggo	
Plant species	**Plant family**	**Plant species**	**Plant family**
Arthocarpus altilis	Moraceae	*Arthocarpus* spp	Moraceae
Cananga odorata	Anacardiaceae	*Cananga* spp	Anacardiaceae
Cananga sp2	Anacardiaceae	*Canarium* spp	Burseraceae
Canarium moluccensis	Burseraceae	*Celtis* sp	Ulmaceae
Canarium sp2	Burseraceae	*Disoxylum* sp	Meliaceae
Celtis latifolia	Ulmaceae	*Dispyros* spp	Ebenaceae
Endospermum diadenum	Euphorbiaceae	*Endospermum diadenum*(Miq.) Airy Shaw	Euphorbiaceae
Eugenia anomala	**Myrtaceae**	*Ficus* spp	Moraceae
Eugenia sp2	**Myrtaceae**	**Myristica papuana**	**Myristicaceae**
Ficus hispida	Moraceae	*Vitex pubescens*	Verbenaceae
Ficus pungens	Moraceae	**Syzygium spp**	**Myrtaceae**
Litsea tuberculata	Lauraceae	*Terminalia* spp	Combretaceae
Myristica fatua	**Myristicaceae**		
Myristica papuana Kunth	**Myristicaceae**		
Syzigium spp	**Myrtaceae**		
Terminalia catappa	Combretaceae		
Terminalia complanata	Combretaceae		
Vitex cofassus	Verbenacea		
Vitex pubescens	Verbenacea		
19 species	**10 Family**	**12 Species**	**11 Family**

Appendix 6. Table of forests size based on type of forest cover of Papua Province (km²)

No.	Forest function Regency	PMF	SMF	PF	PSF	SSF	SF	Total
	Southern mainland							
1	Asmat	2954.96	20.97	1393.87	12303.20	93.12	57.02	16823.10
2	Bovendigul	0	0	18210.60	851.86	321.36	5563.96	24947.78
3	Mappi	554.89	33.36	5213.76	79,75.38	913.00	2626.02	17316.41
4	Merauke	3134.14	80.42	6211.80	2245.96	2109.47	5515.92	19297.71
	Total Southern mainland	6643.99	134.75	31030.03	2337.40	3436.95	13762.90	78385.00
	Total (1)							**78385.00**
	Northern Highlands							
5	Jayawijaya	0	0	6825.06	62.37	0	726.06	7611.49
6	Mimika	2584.38	28.66	12799.80	3880.35	314.63	1190.71	20778.53
7	Paniai	0	0	10021.50	1306.34	117.10	1028.92	12473.90
8	Pegunungan Bintang	0	0	11930.00	154.25	1.64	1014.21	13100.08
9	Puncak Jaya	0	0	5669.64	1947.07	37.95	324.92	7979.58
10	Tolikara	0	0	5704.95	2530.41	206.07	596.32	9037.75
11	Yahukimo	0	0	8810.96	806.18	15.45	2779.70	12412.29
	Total Northern Highlands	2584.38	28.66	61739.90	10687	692.84	7660.84	83393.62
	Total (2)							**83393.62**
	Northern lowland mainland							
12	Jayapura	1.39	0	9542.33	986.71	91.53	2058.81	**12680.77**
13	Keerom	0	0	6999.65	261.79	107.09	1107.17	8475.70
14	Mamberamo Raya	1157.47	462.73	16584.60	5779.48	650.90	1298.21	**25933.35**
15	Nabire	226.06	8.68	7804.60	1051.91	524.08	2590.14	12205.47
16	Sarmi	31.02	0.95	9944.18	2103.82	368.45	1631.63	**14080.05**
17	Jayapura Municipality	2.79	0	414.59	45.25	11.15	263.00	736.78
18	Waropen	262.96	0.46	3424.18	733.70	202.11	263,67	4887.08
	Total Northern lowland mainland	1681.69	472.82	54714,10	10962.70	1955.31	9212,63	**78999.20**
	Total (3)							
	Islands							
19	Biak-Numfor	51.70	1.29	1147.15	4.94	6.08	541.54	1752.70
20	Yapen	7.26	32.61	1823.69	1.42	15.45	395.35	2275.78
21	Supiori	29.61	1.03	1.99	0	0	535.48	**568.11**
	Total islands	88.57	34.93	2972.83	6.36	21.53	1472.37	4596.59
	Total (4)							**4596.59**
	Total (1+2+3+4)							**245374.41**

Notes: This data was adapted from: GIS Laboratory of BPKH X Jayapura (2010)

Appendix 7. Red List Category and Criteria of Victoria Crowned Pigeon (*Goura victoria*)

Vu A2cd + 3cd + 4cd, where:

Vu: Vulnerable

A taxon is **Vulnerable** when the best available evidence indicates that it meets any of the following criteria **A to E,** and it is therefore considered to be facing a high risk of extinction in the wild

A: Reduction population based on any of the following: **1 (a) to (e) until 4**

2: an observed, estimated, inferred or suspected population size reduction of ≥30% over the last to 10 years or three generations, whichever is the longer, where the reduction or its causes may not have ceased OR may not be understood OR may not be reversible, based on (and specifying) any of (**a**) to (**e**) under **A1**

     c: a decline in area occupancy, extent of occurrence and/or quality of habitat

     d: actual or potential levels of exploitation

3: A population size reduction of ≥30%, projected or suspected to be meet within 10 years or three generations, whichever is the longer (up to a maximum of 100 years), based on (and specifying) any of (**b**) to (**e**) under A1

     c: a decline in area occupancy, extent of occurrence and/or quality of habitat

     d: actual or potential levels of exploitation

4: An observed, estimated, inferred, projected or suspected population size reduction of ≥30% over any 10 years or three generation period, whichever is longer (up to a maximum of 100 years in the future), where the time period must include both the past and the future, and where the reduction or its caused may not have ceased OR may not be understood OR may not reversible, based on (any specifying) any of (**a**) to (**e**) under **A1**

     c: a decline in area occupancy, extent of occurrence and/or quality of habitat

     d: actual or potential levels of exploitation

Source : IUCN 2001. *IUCN Red List Categories and Criteria: Version 3.1.*

     IUCN Species Survival Commission. IUCN, Gland, Switzerland and

     Cambridge, UK. ii+30 pp. Available from:

     http://www.iucn.org/themes/ssc/redlists/RLcats2001booklet.html.

     (accessed October 8, 2012).

Appendix 8. Table of abbreviations

BKSDA	:	Balai Konservasi Sumber Daya Alam
BPKH	:	Balai Pemantapan Kawasan Hutan
BPS	:	Badan Pusat Statistik
CI	:	Conservation International
CIFOR	:	Center for International Forestry Research
CITES	:	The Convention on International Trade in Endangered Species of wild Fauna and Flora
CMWMA	:	The Crater Mountain Wildlife Management Area
CP	:	Crowned Pigeon
EC-CITES	:	Europa Commision for CITES
FWI	:	Forest Watch Indonesia
HPK	:	Hutan Produksi yang dapat diKonversi
IUCN	:	International Union for Conservation of Nature and Natural Ressources
Keppres RI	:	Keputusan Presiden Republik Indonesia
LIPI	:	Lembaga Ilmu Pengetahuan Indonesia
Ondoafi	:	Tribe's chef in local Papuan culture
PERDASUS	:	Peraturan Daerah Khusus
PF	:	Primary Forest
PMF	:	Primary Mangrove Forest
PNG	:	Papua New Guinea
POLSUSHUT	:	Polisi Khusus Kehutanan
PSF	:	Primary Swamp Forest
PPRI	:	Peraturan Pemerintah Republik Indonesia
PtPPMA	:	Perkumpulan Terbatas Pengkajian dan Pemberdayaan Masyarakat Adat
SCP	:	Southern Crowned Pigeon
SF	:	Secondary Forest
SSF	:	Secondary Swamp Forest
SMF	:	Secondary Mangrove Forest
UURI	:	Undang-Undang Republik Indonesia
VCP	:	Victoria Crowned Pigeon
WCP	:	Western Crowned Pigeon
WWF	:	World Wildlife Fund

Curriculum Vitae

Name	:	Henderina Josefina Keiluhu
Date/Place of birth	:	March 19, 1962 in Sumbawa Besar, West Nusa Tenggara
Nationality	:	Indonesia
Email address	:	hennylewis@yahoo.com

Education

1969-1974	:	Elementary School at SD YPPK Gembala Baik Abepura, Jayapura
1975-1977		Junior High School at SMP Negeri I Jayapura
1978-1981		Senior High School at SMA N 414 (1) Abepura, Jayapura
07/1982-10/1987	:	Undergraduate at Department of Forestry, Agricultural Faculty, University of Cenderawasih, Manokwari-Papua, Indonesia, Majoring in: Forest Production
09/1997-11/2000	:	Graduate at Study Program of Biology, Magister Program, Faculty of Mathematics and Natural Science, University of Indonesia, Jakarta-Indonesia. Majoring in Conservation Biology. *Thesis title*: Pola produksi serasah dan kecepatan dekomposisi serasah daun *Rhizophora mucronata* di Hutan Mangrove Teluk Yotefa, Jayapura-Papua. *Translation*: Litter production pattern and rate of composition of *Rhizophora mucronata* leave litter in Mangrove Forest of Yotefa Bay, Jayapura-Papua.
10/2008-09/2013	:	Postgraduate at Department of Biological Diversity and Ecology, Faculty of Biology Incl. Psychology. Center for Nature Conservation, Georg-August University of Goettingen, Germany. Dissertation title: Impact of hunting on Victoria Crowned Pigeon (*Goura victoria*: Columbidae) in the rainforest of Northern Papua-Indonesia. Supported by DAAD and Department of Education Papua Province -Indonesia.

Occupation

1989 - 2002	:	Lecturer at Department of Forestry, Agricultural Faculty, University of Cenderawasih, Manokwari-Papua, Indonesia.
2002 - current	:	Lecturer at Department of Biology, Faculty of Mathematics and Natural Science, University of Cenderawasih, Jayapura-Papua, Indonesia.